UV Disinfection and Disinfection By-Product Characteristics of Unfiltered Water

Awwa
Research
Foundation

About the Awwa Research Foundation

The Awwa Research Foundation (AwwaRF) is a member-supported, international, nonprofit organization that sponsors research to enable water utilities, public health agencies, and other professionals to provide safe and affordable drinking water to consumers.

The Foundation's mission is to advance the science of water to improve the quality of life. To achieve this mission, the Foundation sponsors studies on all aspects of drinking water, including supply and resources, treatment, monitoring and analysis, distribution, management, and health effects. Funding for research is provided primarily by subscription payments from approximately 1,000 utilities, consulting firms, and manufacturers in North America and abroad. Additional funding comes from collaborative partnerships with other national and international organizations, allowing for resources to be leveraged, expertise to be shared, and broad-based knowledge to be developed and disseminated. Government funding serves as a third source of research dollars.

From its headquarters in Denver, Colorado, the Foundation's staff directs and supports the efforts of more than 800 volunteers who serve on the board of trustees and various committees. These volunteers represent many facets of the water industry, and contribute their expertise to select and monitor research studies that benefit the entire drinking water community.

The results of research are disseminated through a number of channels, including reports, the Web site, conferences, and periodicals.

For subscribers, the Foundation serves as a cooperative program in which water suppliers unite to pool their resources. By applying Foundation research findings, these water suppliers can save substantial costs and stay on the leading edge of drinking water science and technology. Since its inception, AwwaRF has supplied the water community with more than $300 million in applied research.

More information about the Foundation and how to become a subscriber is available on the Web at **www.awwarf.org**.

UV Disinfection and Disinfection By-Product Characteristics of Unfiltered Water

Prepared by:
Paul C. Wobma
CH2M HILL, Inc., Deerfield Beach, FL 33441

William D. Bellamy
CH2M HILL, Inc., Englewood, CO 80112

James P. Malley
University of New Hampshire, Durham, NH 03824
and

David A. Reckhow
University of Massachusetts, Amherst, MA 01003

Jointly sponsored by:
Awwa Research Foundation
6666 West Quincy Avenue, Denver, CO 80235-3098
and

City of Winnipeg
Winnipeg, Manitoba

Published by:

Awwa Research Foundation
Advancing the Science of Water®

IWA Publishing

DISCLAIMER

This study was funded by the Awwa Research Foundation (AwwaRF). AwwaRF assumes no responsibility for the content of the research study reported in this publication or for the opinions or statements of fact expressed in the report. The mention of trade names for commercial products does not represent or imply the approval or endorsement of AwwaRF. This report is presented solely for informational purposes.

Published by IWA Publishing, Alliance House, 12 Caxton Street, London SW1H 0QS, UK

Telephone: +44 (0) 20 7654 5500; Fax: +44 (0) 20 7654 5555; Email: publications@iwap.co.uk;
Web: **www.iwapublishing.com**

AwwaRF report number 91039F. Originally published by AwwaRF for its subscribers in 2004.
IWA Publishing version published 2005.

ISBN 1 84339 928 8

CONTENTS

TABLES

FIGURES

FOREWORD

The Awwa Research Foundation is a nonprofit corporation that is dedicated to the implementation of a research effort to help utilities respond to regulatory requirements and traditional high-priority concerns of the industry. The research agenda is developed through a process of consultation with subscribers and drinking water professionals. Under the umbrella of a Strategic Research Plan, the Research Advisory Council prioritizes the suggested projects based on current and future needs, applicability, and past work; the recommendations are forwarded to the Board of Trustees for final selection. The foundation also sponsors research projects through the unsolicited proposal process; the Collaborative Research, Research Applications, and Tailored Collaboration programs; and various joint research efforts with organizations such as the U.S. Environmental Protection Agency, the U.S. Bureau of Reclamation, and the Association of California Water Agencies.

This publication is a result of one of these sponsored studies, and it is hoped that its findings will be applied in communities throughout the world. The following report serves not only as a means of communicating the results of the water industry's centralized research program but also as a tool to enlist the further support of the nonmember utilities and individuals. Projects are managed closely from their inception to the final report by the foundation's staff and large cadre of volunteers who willingly contribute their time and expertise. The foundation serves a planning and management function and awards contracts to other institutions such as water utilities, universities, and engineering firms. The funding for this research effort comes primarily from the Subscription Program, through which water utilities subscribe to the research program and make an annual payment proportionate to the volume of water they deliver and consultants and manufacturers subscribe based on their annual billings. The program offers a cost-effective and fair method for funding research in the public interest.

A broad spectrum of water supply issues is addressed by the foundation's research agenda: resources, treatment and operations, distribution and storage, water quality and analysis, toxicology, economics, and management. The ultimate purpose of the coordinated effort is to assist water suppliers to provide the highest possible quality of water economically and reliably.

The true benefits are realized when the results are implemented at the utility level. The foundation's trustees are pleased to offer this publication as a contribution toward that end.

With the advent of ultraviolet (UV) disinfection as a viable potable water disinfection process, it was determined that research was needed to better understand the disinfection and operational impacts that might result when applying UV disinfection to unfiltered supplies. This 14-month study assessed the impact of total organic carbon, turbidity, and algal concentrations on disinfection efficiency and operational considerations. In addition, the study evaluated the changes in disinfection byproducts that can result from changes in oxidation and disinfection processes that might accompany the implementation of UV disinfection. The results of this research provides a good understanding of the impact of water quality aspects on UV disinfection efficacy, operations, and the opportunity to modify oxidation and disinfection practices to better protect public health.

Walter J. Bishop
Chair, Board of Trustees
Awwa Research Foundation

James P. Manwaring, P.E.
Executive Director
Awwa Research Foundation

ACKNOWLEDGMENTS

CH2M HILL would like to acknowledge the efforts of the staff and management of the City of Winnipeg, the participating utility in this study. The continuous UV pilot operation and the numerous testing and studies conducted during the program required a great deal of planning, coordination, and dedication by the City's staff to ensure project success. Special thanks go to Tom Pearson, Mike Shkolny, Duane Griffin, Kelly Kjartanson, Steve Fletcher, Frank DeVries, Glen Chambers, Trevor Hutchinson, Mark Hrynkiw, and Darry Drohomerski.

CH2M HILL would like to acknowledge the other project team participants. Special thanks go to James Malley and Craig Douglas of the University of New Hampshire for their efforts and contributions in performing the UV inactivations studies, and to David Reckhow, Narayan Venkaatesan, Guanghui Hua, and their team at the University of Massachusetts for their efforts and contributions in performing the disinfection and disinfection byproduct studies.

CH2M HILL would also like to acknowledge Albert Ilges, the AwwaRF project manager, for his input and patience throughout the project. Also the Project Advisory Committee – Dan Schmelling (USEPA), Salome Freud (NYC Department Environmental Protection), and Erin Mackey (Carollo Engineers), and the Technical Advisory Committee – Richard Miltner (USEPA), Eva Nieminski (State of Utah), Steve Hubbs and Rengao Song (Louisville Water Company), William Robertson (Health Canada), Margaret Fast (Winnipeg Regional Health Authority), and Don Rocan (Manitoba Conservation).

CH2M HILL also acknowledges the participating UV manufacturers, Aquionics and Wedeco-Ideal Horizons, who provided equipment and technical support during the study.

EXECUTIVE SUMMARY

BACKGROUND AND OBJECTIVES

With the recent recognition of the effectiveness of ultraviolet (UV) disinfection for protozoan cysts and oocysts, interest has increased in the full range of possible drinking-water applications for UV disinfection. Drinking-water studies to date have concentrated on filtered waters with low turbidities and relatively low organic and algae concentrations. By identifying and understanding the impact of turbidity, algae, and organics on UV efficacy, utilities can consider a broader range of UV applications, such as unfiltered systems, bank filtration, prefiltration, and alternative water sources.

In addition to the advantages offered by UV for disinfection, it is possible to significantly alter disinfection and oxidation techniques with a resulting change in disinfection byproduct (DBP) formation and species. Understanding the possible impacts on DBPs, both concentration and type, is essential for comprehensive public health planning. The City of Winnipeg's source water quality is ideal for the study of unfiltered water quality impacts on UV disinfection efficacy and changes in DBPs. With this water source, it was possible to study the effects of turbidity, total organic carbon (TOC), and algae on UV disinfection efficacy and creation/production of DBPs. Turbidity and algae are of particular interest because they can shield pathogens from UV light, resulting in a reduction in UV efficacy.

Based on the uncertainties in UV disinfection resulting from turbidity, algae, and TOC, the following specific objectives were developed for this research project:

- Determine the impact of turbidity on UV disinfection efficacy (pilot- and bench-scale testing)
- Determine the impact of algae on UV disinfection efficacy (pilot- and bench-scale testing)
- Determine the impact of TOC on UV disinfection efficacy (pilot- and bench-scale testing)
- Determine the possible reduction and change in DBPs attainable after converting from free chlorine to UV for primary disinfection (bench-scale testing)
- Determine the impact of algae and TOC on lamp fouling characteristics and cleaning needs over a 1-year period (pilot-scale testing)

The experimental design was developed so that three areas of investigation could be conducted simultaneously with close coordination to ensure quality and verification of results. The areas of investigation included the following activities:

- UV pilot testing conducted continuously over 1 year to determine operation and maintenance requirements such as lamp fouling rates and cleaning requirements.
- UV inactivation studies conducted periodically over 1 year using collimated beam (CB) tests to verify disinfection efficacy, for variable water quality, and to assist in determining operation and maintenance requirements such as lamp fouling rates and cleaning requirements.

- Bench-scale and laboratory studies of DBPs to determine possible DBP control strategies after converting to UV primary disinfection.

UV disinfection pilot testing and disinfection studies, as well as DBP bench testing, were conducted over a 14-month period. The City of Winnipeg constructed the UV pilot plant inside a refurbished construction trailer. Both low-pressure and medium-pressure UV pilot units were included in the pilot facility. Bench-scale tests were conducted in the City's laboratory, as well as at the University of Massachusetts (UMass) and at the University of New Hampshire (UNH).

RESULTS

For the City of Winnipeg's unfiltered water supply, both low- and medium-pressure UV reactors delivered a target UV dose of 40 mJ/cm^2 to the water through seasonal water quality variations. UV dose determinations were best made by calculating reduction equivalent doses through biodosimetry testing.

Turbidity spiking tests were carried out due to a lack of a natural high turbidity event. These tests indicated that the sediments from the source water basin had little to no effect on the log inactivation or reduction equivalent dose and that no significant change in reactor performance was seen at turbidity levels up to 4 nephalometric turbidity units (NTU). At a turbidity level of 13 NTU, there was a reduction in UV disinfection efficacy. It is important to note that there are uncertainties and limitations when spiking turbidity into a system to simulate a turbidity event and also in spiking the challenge organism. The degree to which turbidity spiking matches natural events is beyond the scope of this study and could not be determined. Turbidity spiking results suggest that for unfiltered supplies that meet U.S. Environmental Protection Agency's (USEPA) unfiltered water requirements, UV can provide the required disinfection without compensation for turbidity. However, it is not the intent nor within the scope of this study to imply that UV disinfection is an adequate replacement for proper filtration.

An increase in TOC values resulted from increased algal concentrations and natural changes in the source water matrix. Increasing TOC resulted in decreasing ultraviolet transmittance (UVT). During higher TOC events, UV reactors required higher energy to maintain the desired dose. This effect on the UV reactors is expected to be minor with source waters with a narrow range of TOC values.

There is an inverse relationship between plankton counts and UVT. This suggests that UV reactors will require higher energy during elevated plankton levels in source waters due to the decrease in UVT resulting from the increase in TOC originating from the algae. Once the UV intensity was corrected, the MS-2 phage had the same response to delivered UV dose as the samples where the plankton counts were not high. This confirms that the plankton are not shielding or otherwise protecting the MS-2 biodosimeter from UV inactivation.

UV intensity sensors for both medium-pressure and low-pressure high output reactors were inconsistent and unresponsive to significant changes in performance. For several inactivation studies carried out before and after lamp sleeve cleaning, reduction equivalent doses were seen to increase by as much as 100 percent while sensor values showed little change in intensity levels.

For this source water, UV lamp sleeves required cleaning approximately every 2,000 hours. Lamps and sleeves require replacement after approximately 5,000 hours. of operation. These cleaning and replacement periods will differ for varying source waters due to varying water quality. In addition, the automatic wiper feature on the medium-pressure reactor was disabled so lamp fouling could be compared to the low-pressure high output unit. It is anticipated that lamp sleeve cleaning intervals would be extended for the medium-pressure unit with the automatic wiper feature enabled.

It was found that DBP formation increases with chlorine dose and contact time under all pH conditions. Trihalomethane (THM) concentration increased with pH while HAA5 concentration decreased with pH at all doses and contact times. The trends for the kinetic data for this water were generally in agreement with those reported for other waters in the literature.

DBP testing showed that total THM (TTHM) levels, as produced by the various disinfection scenarios, resulted in concentrations between 25 and 55 μg/L for prechloramination scenarios and 45 to 120 μg/L for those involving prechlorination. Total haloacetic acid (HAA5) levels in the final treated samples ranged from 25 to 55 μg/L when prechloramination was used. The corresponding values for prechlorination trains were 40–100 μg/L. Results suggest that by converting to UV disinfection and using chloramination for the distribution system, DBP goals and regulations can be met.

DBP tests also indicated that there was no obvious difference between the sequences UV/Cl_2 and Cl_2/UV with respect to the THM and HAA formation. However, at higher UV doses (>400 mJ/cm^2), a clear trend indicated lower total organic halides (TOX) and nitrosodimethylamine (NDMA) formation in the Cl_2/UV sequence as compared to the UV/Cl_2 sequence. Also, at the higher UV doses, UV treatment caused a noticeable increase in chlorine demand, as expected, due to the destruction of chlorine by the UV. Between 0 and 400 mJ/cm^2, the increase in demand was about 20 percent. At UV doses greater than 400 mJ/cm^2, DBP concentrations also increased with increasing UV dose. This was especially noticeable with TOX formation.

RECOMMENDATIONS

UV intensity sensor performance bears watching in full-scale UV system selection, design, and operation. It is essential during UV reactor validation and long-term operation that the sensors meet tight specifications initially and are proven periodically to be calibrated and responsive to significant changes in performance (reduction equivalent dose) caused by such common factors as lamp fouling, flow changes, lamp aging, etc. Further sensor research should be conducted to find a sensor(s) that can accurately account for fluctuations in turbidity and organic levels in the source water, as well as lamp fouling.

The effects of UV dose on chlorine demand and DBP formation should be confirmed by utilities implementing UV for primary disinfection. The effects of the sequence of UV and free chlorine should also be confirmed on a site-specific basis. While it is anticipated that these effects will be minimal at typical UV doses used for water treatment (i.e., <60 mJ/cm2), during this study, significant effects on DBPs and chlorine demand were seen at higher UV doses.

It is reasonable to consider UV disinfection for unfiltered systems that comply with USEPA regulatory requirements.

CHAPTER 1
INTRODUCTION

PROJECT BACKGROUND

The development of UV disinfection as a reliable water treatment technology requires a substantial commitment by the water industry. With the recent recognition of the effectiveness of UV disinfection for protozoan cysts and oocysts, interest has increased in the full range of possible drinking-water applications for UV. Potential UV disinfection applications include groundwater, prefiltration, postbank filtration, unfiltered water supplies, open-reservoir treatment, and others.

However, drinking-water studies to date have concentrated on filtered waters with low turbidities and relatively low organic and algae concentrations. By identifying and understanding the impact of turbidity, algae, and organics on UV efficacy, utilities can consider the broader range of UV applications listed above. Partial understanding of the impact of these constituents not only limits the application of UV disinfection but also could lead to an inappropriate application of UV technology. This research is considered to be basic to the full understanding of UV disinfection of potable water.

In addition to the advantages offered by UV for disinfection, it is possible to significantly alter oxidation techniques and DBP formation. Understanding the possible impacts on DBPs, both concentration and type, is essential for comprehensive public health planning. Additional research into possible DBP compliance strategies offered by UV is necessary to complete this picture. The water supply and treatment system of the City of Winnipeg, Manitoba, offers a unique opportunity to investigate a wide range of DBP precursor concentrations and oxidant and disinfectant combinations that are not typically present in filtered-water supplies.

OBJECTIVES

The City of Winnipeg, Manitoba (population 632,000), is implementing UV for primary disinfection as a key component of its water supply and treatment process. UV disinfection offers the following public health benefits:

- Provides an additional barrier against viruses, *Cryptosporidium, Giardia,* and other possible pathogens
- Reduces DBPs

Recent studies have demonstrated the effectiveness of UV disinfection for the inactivation of *Cryptosporidium* oocysts and *Giardia* cysts. The verification of UV disinfection efficiency combined with advances in lamp technology, and increased equipment reliability has made the application of UV a cost-effective alternative for drinking-water applications. However, a number of uncertainties are associated with UV efficacy for waters with variable quality (e.g., the impact of turbidity, algae, and organics). In addition, these site-specific variables can affect other operating conditions, such as lamp fouling rates, UV transmittance, and DBPs.

On the basis of these uncertainties, the following specific objectives were developed for this research project:

- Determine the impact of turbidity on UV disinfection efficacy (pilot- and bench-scale testing)
- Determine the impact of algae on UV disinfection efficacy (pilot- and bench-scale testing)
- Determine the impact of total organic carbon (TOC) on UV disinfection efficacy (pilot- and bench-scale testing)
- Determine the possible reduction and change in DBPs attainable after converting to UV for primary disinfection (bench-scale testing)
- Determine the impact of algae and TOC on lamp fouling characteristics and cleaning needs over a 1-year period (pilot-scale testing)

The results of this study provide information necessary to determine the appropriateness of including UV in the City's treatment plans, as well as provide information to other utilities that are considering UV for situations other than disinfection of filtered waters. This might include systems with groundwater, bank filtration, uncovered reservoirs, unfiltered supplies, and possibly UV before filtration. The DBP information demonstrates possible approaches and anticipated outcomes for changes in the disinfection system to control the type and concentration of DBPs when using UV as the primary disinfectant.

CHAPTER 2
BACKGROUND

THE CITY OF WINNIPEG'S SYSTEM

The City of Winnipeg's (City's) water supply and distribution system is shown in Figure 2.1. The water supply originates in Shoal Lake, which is located on the border between Manitoba, Ontario, and Minnesota. The raw water is chlorinated at the Shoal Lake headworks at typical doses of 3-6 mg/L and flows by gravity through a 137-km (85-mile) aqueduct to the Deacon Reservoir at the east end of the City. The residence time in the aqueduct from the headworks to the Deacon Reservoir is approximately 30 hours. Water for distribution to the City is drawn from the Deacon Reservoir and conveyed to the City's covered reservoirs for distribution to consumers. Water treatment before distribution consists of chlorination, fluoridation, and corrosion control.

For 25 years, the City has been adding chlorine continuously at the Shoal Lake intake to maintain a free chlorine residual throughout the aqueduct to the Deacon Reservoir. The reasons for continuous chlorination of the aqueduct are as follows:

- Maintain a barrier against potential contamination of the drinking-water supply by microbial pathogens
- Control the growth of slime within the aqueduct, which could reduce the hydraulic capacity
- Prevent zebra mussels from entering the aqueduct and attaching to the interior surfaces, which would reduce the hydraulic capacity of the aqueduct
- Assist in controlling taste and odor

Unfortunately, free chlorination of organics that naturally occur in the water of Shoal Lake produces DBPs that may pose a chronic health risk to consumers. In addition, chlorine disinfection may not provide sufficient protection against all pathogens of concern. The conflict between the objectives of continuous chlorination and the need to minimize the production of DBPs to mitigate the potential chronic effect on public health provided the impetus for this study.

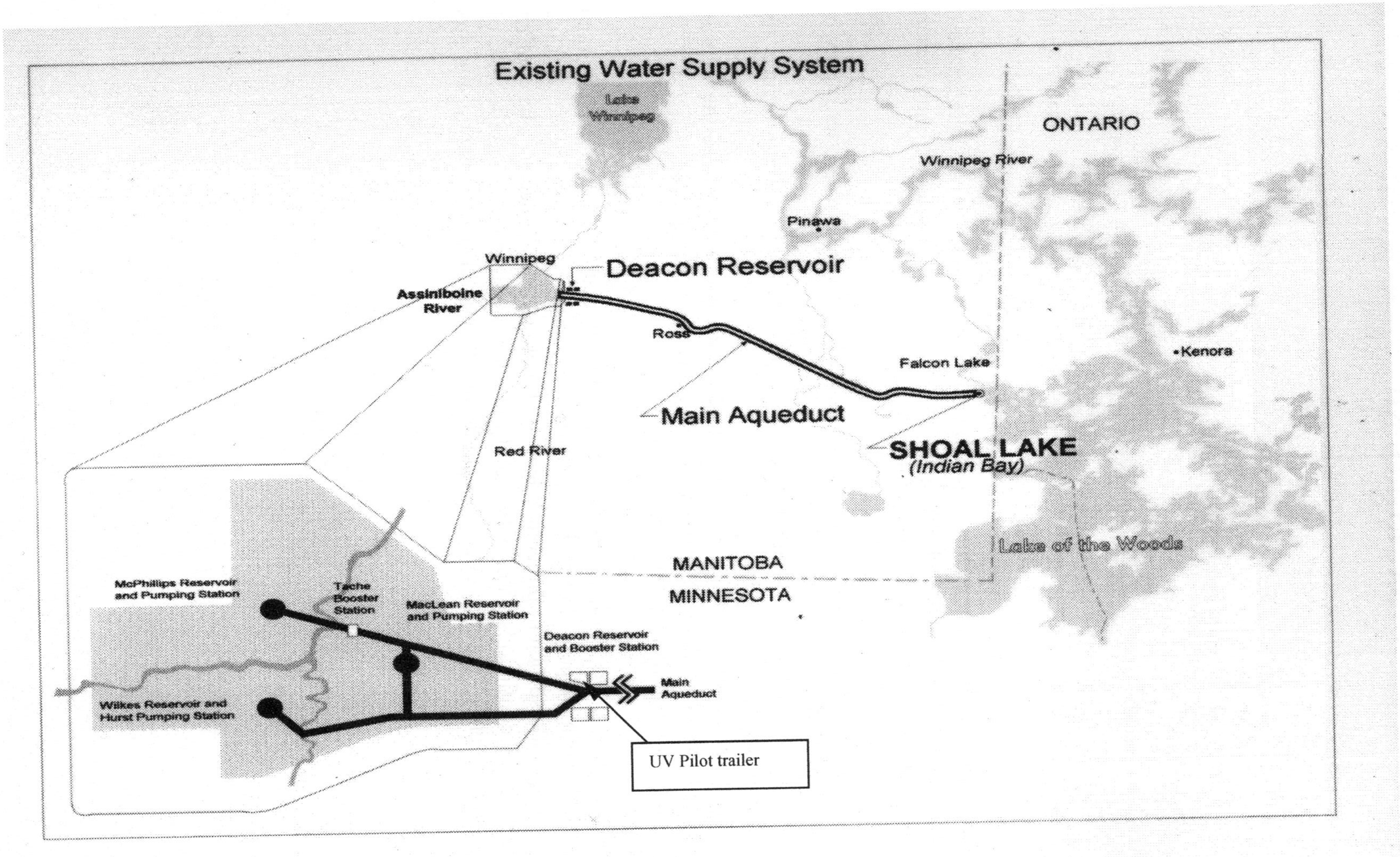

Figure 2.1 City of Winnipeg water supply and distribution

BASIS FOR THE STUDY

The City's source water quality is ideal for the study of unfiltered water quality impacts on UV disinfection efficacy and changes in DBPs. Turbidity, TOC, and algae can and do interfere with disinfection efficacy. TOC and algae byproducts influence the concentration and speciation of DBPs. Historical raw water quality data for Deacon Reservoir are presented in Table 2.1. Turbidity and algae are of particular interest because they can shield pathogens from UV light, resulting in a reduction in UV efficacy. In addition, the effects of UV on medium- to high-TOC waters have not been investigated in great detail. There are seasonal effects on water quality parameters. During the pilot testing, the water temperature ranged from 0.5 to 25°C. Turbidity ranged from 0.4 to 1.7 NTU with the higher levels occurring in the spring and lower levels in the winter. Similarly, UVT levels ranged from 72.5 to 79.8 percent with lower UVTs in the spring and higher UVTs in the winter. Plankton levels ranged from 1,375 to 285,000 counts/mL; however, the average level was 17,140 counts/mL. Algae blooms typically occurred in late summer.

Table 2.1
Deacon Reservoir historical inlet raw water quality (1989–1994)

Parameter	Units	Average	Minimum	Maximum
pH	-	8.2	7.4	9.1
Turbidity	NTU	1.0	0.3	5.3
TOC	mg/L	9.3	5	17
Dissolved organic carbon	mg/L	8.3	4	15
Plankton	cells/mL	39,700	200	666,000
Alkalinity (total)	mg/L as $CaCO_3$	81	72	95
Hardness (total)	mg/L as $CaCO_3$	83	68	97
TTHMs	µg/L	122	87	185
Color (true)	TCU	<5	<5	10

Drinking-water regulations in the United States and Canada are moving toward stricter requirements for pathogen and DBP control. The USEPA is proposing the Long Term 2 Enhanced Surface Water Treatment Rule (LT2ESWTR) to reduce disease incidence associated with *Cryptosporidium* and other pathogenic microorganisms in drinking water. The LT2ESWTR will supplement existing regulations by targeting additional *Cryptosporidium* treatment requirements to higher risk systems. The LT2ESWTR will apply to all systems that use surface water or groundwater under the direct influence of surface water.

The USEPA has recognized that UV disinfection is a low-cost barrier for *Cryptosporidium* and *Giardia*. However, regulatory consensus on the use of UV disinfection has not been reached in the United States or Canada. The frameworks for testing and verification have been tentatively developed but have not been generally accepted. This research assists in providing the information needed to establish testing protocols and application limits for UV in situations other than disinfection of filtered waters.

CHAPTER 3
METHODOLOGY

SCOPE OF RESEARCH

The experimental design was developed so that three areas of investigation could be conducted simultaneously with close coordination to ensure quality and verification of results. The studies included the following activities:

- UV pilot testing continuously over 1 year to determine operation and maintenance requirements such as lamp fouling rates and cleaning requirements
- UV inactivation studies periodically over 1 year to verify collimated beam (CB) tests of disinfection efficacy, for variable water quality, and to assist in determining operation and maintenance requirements such as lamp fouling rates and cleaning requirements
- Bench-scale and laboratory studies of DBPs to determine possible control strategies for DBPs after converting to UV primary disinfection

PILOT TESTING

The City constructed the UV pilot plant inside a refurbished construction trailer. The following describes the pilot- and bench-scale methods and procedures.

Raw Water Supply

The raw supply for the pilot plant came from Deacon Reservoir, which comprised four cells. The pilot plant was located adjacent to Cell Number (No.) 1. Raw water was drawn from Cell No. 1 and discharged back into the reservoir. Water from Cell No. 1 was chlorinated before distribution.

Process Trains

Two UV disinfection units were evaluated:

- Aquionics: Model MD150F/2/3 – medium-pressure, 100 mm diameter, 1 lamp, 1 sensor
- Wedeco-Ideal Horizons: Model LBA20 – low-pressure-high-intensity, 100 mm diameter, 4 lamps, 1 sensor

These UV units were piloted in parallel to compare treatment performance on similar water and to develop operation and maintenance cost parameters. With respect to lamp cleaning, the Aquionics unit has an automatic wiper feature for periodic cleaning of the lamp sleeve. The Wedeco unit did not have wipers on the sleeves but uses periodic in place cleaning with a weak acid to remove any films or deposits on the lamp sleeve. To standardize the lamp cleaning investigation, the automated wiper feature on the Aquionics unit was turned off for the pilot program. Both units were hand cleaned as required using the same procedure.

Pilot Plant

The pilot plant process flow diagram is presented in Figure 3.1. The raw water pump provided water to both UV trains in parallel. The common raw water line had three injection points upstream of a static mixer. The first injection point was for free chlorine injection. The second was for an alternate chemical. The third injection point was for the injection of MS-2 coliphage, which served as the biodosimeter.

Each UV train included flow meters, control valves, UV reactor, pressure gauges upstream and downstream from the reactor, sample ports upstream and downstream from the reactor, and a downstream static mixer. Each UV reactor included a control panel, a chemical cleaning tank, and an online transmittance sensor. The UV manufacturer representatives were onsite for startup and training services. The two UV pilot units are shown in photographs as Figures 3.2 and 3.3.

Test Period

The pilot test period covered approximately 14 months of continuous operation. The intention was that the pilot plant would operate continuously for 12 months; however, it was actually in operation for approximately 14 months to allow for specific tests that were deemed appropriate for this study. The objective during this period was to determine the operating and maintenance characteristics associated with lamp fouling and the variability in UV dosage and light intensity due to changes in water quality and lamp aging.

Five inactivation studies were included during the pilot program. The intention was to carry out the inactivation studies during the following water quality periods:

- Average turbidity, algae, and TOC
- High TOC
- High algae
- High turbidity or repeat session with high algae

These water quality periods were not necessarily unique or well defined. Therefore, testing times were selected based on seasonal water quality and during raw water conditions that most closely matched the desired quality. Bench-scale testing was conducted concurrent with the pilot plant operation and included CB studies and disinfection and DBP testing.

Steady State Pilot Operations

The City provided the staff to operate the pilot facility. The system was operated without any prechlorination (other than that already provided at the Shoal Lake Intake).

In addition to operating the pilot plant, the City carried out the following routine analyses:

- Recorded raw water flow, temperature, and pressure for each stream
- Recorded turbidity, particle counts, pH, temperature (bench)
- Tested TOC, alkalinity, hardness, iron, manganese, UV254 (lab)
- Tested plankton counts, chlorophyll a, simulated distribution system DBPs (lab)
- Recorded lamp operating time
- Recorded UV intensity, UV dose, cleaning frequency

Table 3.1 presents the pilot test parameters and frequency.

Using the manufacturer's standard calibration procedures, City personnel also calibrated and cleaned the UV units. Calibration of dose was conducted after each inactivation study. Lamp sleeves and lamps were cleaned during each inactivation study to determine the impact of sleeve fouling. Figures 3.4, 3.5, 3.6, and 3.7 illustrate the cleaning and calibration that occurred during the study.

Table 3.1
Water quality sampling frequency and locations – steady state operations

Parameter	Sample/monitor frequency	Sample/monitor location	Analytical method
Standard			
Flow	M, W, F	Each UV stream	Local indicator
Temperature	M, W, F	Pilot raw water	Local indicator
Pressure	M, W, F	Upstream and downstream of each UV reactor	Local indicator
Turbidity	M, W, F	Pilot raw water	Bench unit
Particle counts	M, W, F	Pilot raw water	Bench unit
pH	M, W, F	Pilot Raw water	
Lamp operating time	M, W, F	Each UV reactor	Bench unit
UV intensity	M, W, F	Each UV reactor	Local indicator
UV power consumption	M, W, F	Each UV reactor	Local indicator
UV dose	M, W, F	Calculated (use intensity and flow)	N/A
Analytical			Standard method
UV254	Weekly	Pilot raw water	5910 B (unfiltered) 19th Ed., 1995
TOC	Weekly	Pilot raw water	5310 B 19th Ed., 1995
Chlorine residual	Weekly	Pilot raw water	4500-Cl D, 4500-Cl E 19th Ed., 1995
Plankton counts	Every 2 weeks	Pilot raw water	1002 A – F 17th Ed., 1985
Chlorophyll a	Every 2 weeks	Pilot raw water	10200 H 19th Ed., 1995

(continued)

Table 3.1 (continued)

Parameter	Sample/monitor frequency	Sample/monitor location	Analytical method
Analytical			Standard method
Alkalinity	Weekly	Pilot raw water	2320 B 20th Ed., 1998
Total iron	Weekly	Pilot raw water	3111 B 19th Ed., 1995
Total manganese	Weekly	Pilot raw water	3111 B 19th Ed., 1995
Hardness	Weekly	Pilot raw water	2340 C 20th Ed., 1998
TTHMs	Monthly	Upstream and downstream of each UV reactor (triplicate sets)	6230 C 19th Ed., 1995
THAAs	Monthly	Upstream and downstream of each UV reactor (triplicate sets)	6251 A-B 20th Ed., 1998
Special testing			
Inactivation studies	5 times (based on raw water quality)	Upstream and downstream of each UV reactor (triplicate sets)	UNH
CB studies	5 times (based on raw water quality)	Upstream and downstream of each UV reactor (triplicate sets)	UNH
DBP testing	8 times	Shoal Lake raw water, pilot raw water	UMass

Turbidity Spiking

During the test period, turbidity did not reach the desired 3 to 5 NTU. As a result, it was necessary to collect sediments from the bottom of Deacon Reservoir and spike the feed water to the UV disinfection units. Figures 3.8 and 3.9 show the collection of the spiking material. As shown, the sediment was collected during the winter. A hole was cut into the ice covering Deacon Reservoir, and a diver used a hose and pump to collect sediment. The sediment was subsequently mixed in a tank and injected into the main pilot plant flow at the flow rate necessary to achieve the desired turbidity.

INACTIVATION STUDIES

Introduction

UNH used MS-2 biodosimetry to perform the inactivation studies. The purpose of the MS-2 bacteriophage biodosimetry was to evaluate the UV dose delivered by the Wedeco and Aquionics units as a means of demonstrating system performance. Initially, four inactivation studies were scheduled during the pilot program. However, a fifth inactivation study was added to further evaluate the effects of turbidity on UV performance.

Testing Procedures

Experimental testing was broken into three general categories: UV scans, CB studies, and MS-2 field challenges. Inactivation tests were carried out in June, August, and October 2001, and in March and July 2002. Testing events were selected based on seasonal raw water quality variations.

UV Scan

Objective

UV scans are useful in characterizing the UV absorbance of the water to be studied and tracking how it changes with seasons and/or with changes in weather such as wind events. The UV scans were performed whenever CB and reactor challenge testing was performed. The UV scan data was also used in the CB dose calculations as described below. UV scans, as opposed to measurements solely at 254 nm, have aided utilities in assessing the effectiveness of medium-pressure UV systems. This is in part due to the polychromatic nature of medium-pressure lamp outputs.

Materials and Methods

For the UV scans, the absorbance of each sample was measured using matched 1-cm path length Suprasil Quartz cuvettes in a double-beamed Hitachi 5000 UV-VIS spectrophotometer. The baseline was set using high-quality reverse osmosis (RO) water (Type II Laboratory water). The UV scan was performed in duplicate with absorbance measured at 1 nm increments from 190 nm to 400 nm. The 1 nm increments were chosen because they represent a convenient and reproducible methodology for the Hitachi spectrophotometer. Choosing smaller increments does not significantly increase the usefulness of these data but results in much longer analytical time and much more data storage, retrieval and processing efforts. Absorbance (A) measurements can be converted to percent UVT using the following equation:

$$\%UVT = 10^{-A} \times 100\%$$

CB Study

Objective

The purpose of the CB trials was to determine the inactivation of MS-2 bacteriophage in a laboratory sample of Deacon Reservoir water. The CB data is used to develop a standard curve relating log inactivation to delivered UV dose in mJ/cm^2. This standard curve is then used in the field biodosimetry testing (field challenges) to calculate the reduction equivalent dose (RED) in mJ/cm^2 of the reactor being testing. Results from bench-scale CB testing are a critical part of the MS-2 biodosimetry testing of the pilot units.

Materials and Methods

Sample Preparation. The Deacon Reservoir water was used as the water matrix for the CB studies. For each test, 1.0 mL of MS-2 bacteriophage (~1 x 10^{10} pfu/mL) was added to 1,000 mL of Deacon Reservoir water.

CB Procedures. The following step-wise batch, bench-scale CB procedure was used:

1. Set up equipment according to manufacturer's specifications. (An Infilco Degremont CB unit was used in these studies.)
2. Warm up the lamps for a minimum of 60 minutes to stabilize the UV output before testing. (UV lamps had been previously burned in and were stable.)
3. Measure the UV intensity at multiple points within the area of the CB at the mid-liquid level of the petri dish (the future location of the test sample) using a precalibrated and factory-certified IL-1700 research-grade radiometer (International Light, Inc., Newburyport, Mass.).
4. Place 15 mL of the MS-2–spiked test water onto the stir plate at the same level at which UV intensity was measured.
5. Sample 100 L to determine initial organism concentration.
6. Gently stir and irradiate the sample for required time by opening the UV shutter on the CB unit to produce the desired UV dose.
7. Close UV shutter after predetermined contact time (see below),and remove samples.
8. Verify UV irradiance with radiometer to determine if a significant change has occurred since the onset of testing.
9. Sample to determine final organism concentration in the petri dish.
10. Repeat Step Nos. 3 through 9 for other UV doses.
11. Run duplicate samples to verify precision. Enumerate test organisms. A precision of ± 0.2 log units is considered acceptable, and the data can be used for RED determination.

Irradiation/Dose Calculation. Given the lamp irradiance, a spreadsheet created by Jim Bolton of Bolton Photosciences, Inc. (spreadsheets and discussion are available on the website www.iuva.org) was used to calculate the time needed to achieve average doses of 0, 20, 40, 60, 80, and 100 mJ/cm^2. When calculating the time needed, the spreadsheet accounts for the distribution of the incident monochromatic UV light across the water surface, the absorbance of the water matrix, the germicidal effectiveness of 253.7 nm, the reflection of this monochromatic UV light off the water surface, and the irradiance output of the lamp. Because the USEPA UV Design Guidance Manual (UVDGM) draft was being created over a 3-year period, much discussion and a better understanding was developed on the mathematics of curve fitting UV dose-response curves intended for the development of RED estimates. As a result, the approach for this research project was modified to exclude the 100 mJ/cm^2 values for a more accurate curve fit. The 100 mJ/cm^2 data points were removed because they significantly affect the slopes of the regression lines and the linearity of statistical fit to the data (r^2 value). Because this dose was not representative of the doses being delivered by the pilot reactors, its elimination does not cause any conflicts where the calculated RED values would exceed the CB calibration range. All of the REDs reported in this study have been calculated in accordance with the recommendations discussed in the June 2003 Draft of the USEPA UVDGM.

MS-2 Assay Method. Enumeration of MS-2 was accomplished with the single agar overlay plaque assay (Adams, 1959). After 12 to 16 hours, the number of plaques present on each plate was counted. Plates having fewer than 30 or more than 300 plaques were reported as too few to count (TFTC) or too numerous to count (TNTC), respectively.

MS-2 Field Challenges

Objective

Field challenges constitute the second part of biodosimetry testing. The field challenges combined with the CB studies allowed the calculation of a reliable UV dose for the UV pilot equipment. This dose is often referred to as the RED in the literature because it is derived from a known reduction (inactivation) of the biodosimetry test organism, which, in this case, was MS-2 bacteriophage virus. The RED was used in this UV project to track the performance of the UV system through varying influent water qualities. A CB study was run each time a field challenge was performed. The CB was run within 24 hours at the UNH laboratories.

Materials and Methods

The following list details the general procedure used in the field challenges:

1. Ship approximately 6 L of approximately 10^{10} pfu/ml MS-2 stock to the site. (MS-2 was used within 4 weeks of preparation in all cases.)
2. Label all MS-2 sampling vials (sterile 15 mL polypropylene centrifuge tubes).
3. Record the sample number and sample description on MS-2 challenge data sheet.
4. Verify that there is no free chlorine in the influent water.
5. Ensure both UV systems are stable and running at the prescribed operating conditions.
6. Calibrate the MS-2 feed pump to 100 mL/min ± 2 percent. A continuous feed of MS-2 was used during the challenge studies.
7. Record stock MS-2 concentration reported on the batch shipped from UNH.
8. Take a 15 mL sample of the MS-2 stock for shipment to UNH as a quality assurance/quality control (QA/QC) check. Sample was stored in a sterile 15 mL centrifuge tube and shipped back to UNH on ice for assay to ensure titer was

consistent with titer results from when the batch was shipped to Winnipeg for challenge testing (a standard trip blank essentially).

9. Mix the MS-2 stock with equal parts of unchlorinated influent water making up a minimum of 7.5 L.
10. Allow this spiking mixture to mix for at least 30 minutes before using.
11. Take a 15 mL sample of the spiking mixture for shipment to UNH as a QA/QC check.
12. Take a 15 mL sample of the unspiked influent water to the pilots for shipment to UNH as a QA/QC check.
13. Collect a 1 L sample of unspiked influent water in a brown glass bottle for shipment to UNH for CB studies.
14. Adjust pilot system flow valves to ensure continuous flow from the effluent sampling taps.
15. Adjust pilot system flows to provide the desired rates.
16. Ensure all UV pilot system data are recorded and systems are stable.
17. Start the spiking mixture feed and record initial spike titer, spiking pump flowrate and time on the MS-2 challenge data sheet.
18. Allow spiking mixture to run for 15 minutes before sampling.
19. Take three rounds of influent and effluent samples. The rounds of samples were spaced 15 minutes apart for uniformity and convenience.
20. Record end of run information for both UV pilot systems.
21. Shut off UV lamps on both systems; be sure to keep MS-2 spike running.
22. Sample effluent locations from each UV system after 15 minutes as the final QA/QC check.
23. Turn off spike pump and flush spiking system with influent water.
24. Thoroughly clean all spiking containers, pump, tubing, and sampling location tubing in preparation for next run.
25. Pack all 15 mL tube samples and the 1 L sample carefully to avoid breakage. Add ice packs and ship for next day delivery.

DISINFECTION / DISINFECTION BYPRODUCTS STUDIES

General Experimental Design

Introduction

UMass investigated the impacts of changes in the City's oxidation/disinfection scheme on DBP formation. The investigation included the possible use of chlorine dioxide, chloramines, or intermittent chlorination predisinfection/oxidation; the possible use of UV for primary disinfection (with free chlorine for viral inactivation); and the use of chloramines for distribution system residual.

Aside from the three predisinfectant options (free chlorine, chloramines, chlorine dioxide), UMass also considered three chloramination options (prechlorination, preammoniation, and application of preformed chloramines) and two placements of postchlorination (before and after UV). UMass also examined water quality effects that are likely to impact regrowth in the distribution system.

Materials and Methods

Bench-scale testing for DBP formation was conducted in the UMass laboratories. Raw water samples of approximately 40 L were collected 8 times (June 8, July 16, October 10, November 26 in 2001 and January 30, March 18, June 5, and August 13 in 2002) throughout the project and shipped to UMass for testing. Dates were selected to minimize holding times, while accommodating the Utility's schedule and mindful of the research team's interest in capturing seasonal variations. The intent was to collect samples to represent average conditions and high-TOC, high-algae, and high-turbidity conditions.

Samples of approximately 5 L were treated in a series of batch and flow-through treatment systems aimed at simulating the various pre- and postdisinfection and oxidation options (see Figure 3.10). The preoxidation options included application of the following:

- Preformed chloramines (to minimize DBP formation)
- Ammonia followed by addition of chlorine (a more conventional chloramination practice)
- Chlorination (to simulate a shock or intermittent prechlorination scenario)
- Chlorine dioxide

Oxidant/disinfectant doses were selected to achieve a measurable residual at the end of 24 hours. These doses were decided upon using preliminary bench-scale demand tests.

Two contact times were assessed. These times were selected to represent the range in detention times likely to be encountered in the City's conveyance systems (e.g., 8 days and 30 days). At the start of each experiment and at the end of each contact time, samples were collected for analysis of oxidant residual and DBPs (locations numbers 1 and 5–7 in Figure 3.10). The measured DBPs include the total trihalomethanes (TTHM) (and other neutral extractables) and all nine haloacetic acids (HAA9). Samples were measured in duplicate using USEPA Method 552.2 with only minor modifications.

Before embarking on a full examination of chlorine dioxide as a potential predisinfectant, some simple bench-scale demand studies were run. These tests were to determine the stability of chlorine dioxide in the Shoal Lake raw water and the potential for inorganic byproduct formation (chlorite and chlorate). If these tests showed that chlorine dioxide residuals could be maintained without producing chlorite levels well above 1 mg/L (i.e., 1.2 mg/L), then chlorine dioxide testing would proceed with more detailed bench-scale investigations.

Each of the four pretreatment options was combined with three possible post-treatment scenarios (Figure 3.11). The following combinations were included:

- Free chlorination for virus control followed by UV and application of preformed chloramines for *Cryptosporidium* control and maintenance of a residual in the distribution system
- Free chlorination for virus control followed by UV and application of ammonia and then chlorine for *Cryptosporidium* control and maintenance of a residual in the distribution system
- UV treatment followed by a short free-chlorine contact and then postammoniation for virus and *Cryptosporidium* control and maintenance of a residual in the distribution system

Combined, this test plan resulted in 12 overall treatment permutations. As presented in Figure 3.11, samples were collected immediately after treatment (Nos. 14, 16, 18 for Cl_2/UV and Nos. 15, 17, 19 for UV/Cl_2) and after post-treatment (Nos. 21 through 29). The final samples (Nos. 21 through 29) were collected after a 5-day contact time with the residual disinfectants. This was intended to simulate an extreme distribution system residence time. As with the pretreatment analyses, all samples were measured for residual disinfectant, as well as THMs and HAAs.

UV treatment was conducted at UMass in a 2-L flow-through reactor using a monochromatic LP lamp. The equipment was loaned to UMass from UNH for use throughout the project. Dr. James Malley and his UNH team trained the UMass research assistants to use the equipment.

Testing for Biodegradable Organic Matter

Selected samples, representing the distribution system, were analyzed for biodegradable organic matter (BOM). This included three types of measurements: (1) specific biodegradable compounds (aldehydes and oxo-acids), (2) biodegradable dissolved organic carbon (BDOC), and (3) assimilable organic carbon (AOC). The reason for making all three measurements was that each offers its own unique advantages, and no single measurement has been acknowledged as definitive with regard to regrowth potential (a complete discussion of this issue is beyond the scope of this report; for more information, the reader should consult Xie and Reckhow (1992), Kaplan et al. (1993), Volk et al. (1994), Schechter and Singer (1995), Escobar and Randall (2001), Paode et al. (1997), Volk and Lechevallier (2002), Frias et al. (1995), Allgeier et al. (1996), Huck et al. (1990), Carlson and Amy (2000)).

Bench-Scale Treatment Protocols

Chlorination/Chloramination Procedures

Chlorination and chloramination were conducted through widely used methodology. Generally, reagents were added in the form of concentrated solutions under conditions of high-speed mixing. Symons and co-workers have concluded that the exact nature of this mixing is not of primary importance in simulating full-scale chloramination with bench-scale experiments. Nevertheless, to avoid any possible complication of this type, preformed chloramines were also used. These are produced by carefully mixing concentrated solutions of sodium hypochlorite and ammonium chloride. This is done at low temperatures and at a controlled pH. Experience with this approach at UMass has shown that relatively stable and pure solutions of monochloramine can be produced in this way.

Ozonation Procedures

When required, samples were ozonated in a semi-batch system. Ozone is generated from pure oxygen by means of a laboratory corona discharge generator. The ozone/oxygen product gas is introduced into a 2-L glass reaction vessel containing the water to be treated. Flow is controlled with an electronic flow controller, and the ozone content is monitored by direct UV absorbance spectrophotometry. The gas is mixed with the sample by a porous quartz frit. Off-gas is re-directed through a spectrophotomer for determination of ozone content. A membrane ozone electrode (Orbisphere) is fitted into the side of the glass reactor so that aqueous ozone concentration can be monitored continuously. Ozone transferred is determined from the flow rates and the differences in ozone content in the applied gas versus the off-gas.

Chlorine Dioxide Treatment

When required, samples were preoxidized with chlorine dioxide in a batch reactor. The reaction was conducted at darkness in biological oxygen demand (BOD) bottles in absence of air in order to avoid the possible loss of oxidant or volatile by-products produced during the course of the reaction (flasks were filled up).

Chlorine dioxide was freshly generated as needed. Aqueous solutions were prepared from the gaseous chlorine dioxide generated from the acidification (i.e., sulfuric acid) of a solution of sodium chlorite. To avoid the presence of trace chlorine in the chloride dioxide stock solution, chlorine was removed from the gas stream by a $NaClO_2$ scrubber. Concentrations in chlorine dioxide of solutions prepared using this protocol were found to range from 3 to 4 g/L of ClO_2. The concentration of the chlorine dioxide stock solution was checked before each use using the laboratory solvent blank (LSB) method as developed by Bubnis and others.

Chlorine dioxide typically reacts with most reducing agents through a one-electron transfer; thus, chlorite is considered to be the principal oxidation byproducts and generally represents 50 to 70 percent of the initial chloride dioxide. Based on this and due to the maximum contaminant level (MCL) and maximum residual disinfection level (MRDL) established for chlorite, the practical upper limit for chloride dioxide would be approximately 1.5 mg/L.

Analytical Methods

Total Organic Carbon

TOC was measured by the high-temperature combustion method using a Shimadzu 5000 analyzer.

Residual Chlorine (Free and Combined)

Residual chlorine was measured by titrimetric DPD methodology (4500-Cl, D and F).

THMs and Other Neutral Extractables

THMs and other neutral extractables (haloacetonitriles, haloketones, chloropicrin, etc.) were measured using the standard micro-extraction method with gas chromatograph (GC) and electron capture detection (ECD) (USEPA Method 551.1).

Haloacetic Acids

The full suite of haloacetic acids were measured along with the THMs by the micro-extraction method with methylation and separation/detection by GC with ECD. More specifically, the acidic methanol derivatization (USEPA Method 552.2) was used, which avoids the use of highly toxic reagents as required for the diazomethane method. Acidic methanol has proven to give better and more reliable recoveries of all HAA9 species, especially the brominated forms (Fair, 2000).

Total Organic Halide

TOX was measured on a Euroglass instrument (a Dohrmann DX-20 unit is also available, if needed). Methodology generally follows that established in Standard Methods.

Specific UV Absorbance

The full UV-Visible absorbance spectrum was measured. UV spectroscopy has been extensively used in studying humic substances. Specific UV absorbance at 254 nm is widely used to assess the humic content of natural organic matter (NOM). All absorbance measurements were made at UMass, using a Hewlett-Packard diode array spectrophotometer.

Standard Experimental Protocol

1. Collect initial raw water sample for general analysis

Objective. To establish raw water levels for key parameters
Protocol. Analyze raw water subsample for:

- THMs
- HAAs
- TOC
- UV absorbance
- Ammonia-N

2. Chlorine Dioxide Stability Test

Objective. To determine if ClO_2 can be used in the raw water aqueduct and still meet 1-mg/L chlorite standard
Protocol. Add 1.5 mg/L ClO_2 and test residual over time.
Criteria for continuation of ClO_2 testing:

> 0.1	mg/L residual
at	
30	hours contact time
Ambient	°C
Ambient	pH

3. General Preoxidant Dose Tests

Objective. To determine appropriate preoxidant dose for subsequent tests

Protocol. Add varying doses of preoxidant and test residual at a single characteristic contact time.

- Three preoxidants (chlorine, in situ chloramine, preformed chloramine)
- Six dose levels each (e.g., 2.0, 3.0, 4.0, 5.0, 6.0, 7.0 for free chlorine; 0.5, 1.0, 1.5, 2.0, 3.0, 4.0 for combined chlorine)
- In situ chloramination protocol (with preammoniation)
  - NH_4Cl then NaOCl with intense mixing
  - Cl_2/N ratio: 4 g/g or 0.8 M/M
- Preformed chloramination protocol
  - Prepare stock offline: to 500 mg/L NaOCl solution, add requisite volume of 500 mg/L NH_4Cl solution, with intense mixing; keep on ice; buffer at pH 5 (this pH is best for producing a stable monochlroamine stock)
  - Cl_2/N ratio: 4 g/g or 0.8 M/M
- Monitor chlorine species in stock; add requisite amount to raw water once concentrations have stabilized

Criteria for selection of preoxidant dose for subsequent tests:

0.2	mg/L residual
at	
30	hours contact time
Ambient	°C
Ambient	pH

4. Prechlorination Treatment

Objective. To simulate use of shock prechlorination at the Shoal Lake headworks

Protocol.

- Dose two, 2-L samples of raw water with chlorine at the level determined in Step No. 3 (different for each water); hold for characteristic system contact time (30 hours)
- Two contact times: 8 days and 20 days
- Headspace-free conditions with no attempt to simulate Deacon volatilization or biodegradation

8 and 20	day contact time
Ambient	°C
Ambient	pH

5. Prechloramination Treatment (sequential)

Objective. To simulate use of conventional chloramination (with preammoniation) at the Shoal Lake headworks

Protocol.

- Dose two, 2-L samples of raw water with ammonia and chlorine at the level determined in Step No. 3 (different for each water); hold for characteristic system contact time
- Two contact times: 8 days and 20 days
- Headspace-free conditions with no attempt to simulate Deacon volatilization or biodegradation

8 and 20	day contact time
Ambient	°C
Ambient	pH

6. Prechloramination Treatment (preformed)

Objective. To simulate use of chloramination at the Shoal Lake headworks where order of addition (chlorine/UV versus UV/chlorine) is best suited to minimized DBPs

Protocol.

- Dose two, 2-L samples of raw water with preformed chloramines at the level determined in Step No. 3; hold for characteristic system contact time
- Two contact times: 8 days and 20 days
- Headspace-free conditions with no attempt to simulate Deacon volatilization or biodegradation

8 and 20	day contact time
Ambient	°C
Ambient	pH

7. Prechlorine Dioxide Treatment

Purpose. To simulate use of $C1O_2$ treatment at the Shoal Lake headworks (performed only when criteria in Step No. 2 were met)

Protocol.

- Dose two, 2-L samples of raw water with $C1O_2$; hold for characteristic system contact time (30 hours)
- Two contact times: 8 days and 20 days
- Headspace-free conditions with no attempt to simulate Deacon volatilization or biodegradation

1.5	mg/L dose
8 and 20	day contact time
Ambient	°C
Ambient	pH

8. Post-treatment (repeated for each pretreatment scenario and contact time)

- Open pretreatment reactor (2-L bottle)
- Collect 250-mL sample for general DBP analysis
- Remove 1 L for pre-Cl_2 post-treatment (series A)
- Chlorinate

EPA Table*	mg/L dose
20	minute contact time
Ambient	°C
Ambient	pH

- Check residual
- Pass remaining volume (~750 mL) through UV reactor

40	mJ/cm^2 dose

- Measure chlorine residual
- Transfer to two 300-mL bottles
- Dose one bottle with NH_4Cl to convert free-residual chlorine (FRC) to combined-residual chlorine (CRC) (4 g/g ratio) and then with in situ chloramines (preammoniation)

TBD†	mg/L dose
4	g/g Cl_2/N ratio
5	day contact time
Ambient	°C
Ambient	pH

- Dose second bottle with preformed chloramines

TBD	mg/L dose
4	g/g Cl_2/N ratio
5	day contact time
Ambient	°C
Ambient	pH

---

*Determined from the contact time and the requisite contact time (CT) (from U.S. Environmental Protection Agency CT tables for 2-log virus inactivation).

†Dose to be determined; goal is to have 0.6-mg/L residual at the end of 5 days.

- Open both bottles after 3 days
  - (1) Analyze for general DBPs
  - (2) Analyze for carbonyl compounds
  - (3) Analyze by BOM assays (only 2 of 8 runs)
- Take remaining 750 mL and use for pre-UV post-treatment (series B)
  - (1) Chlorinate; same as with 8c1
  - (2) Sacrifice one bottle and use for general DBP analysis
  - (3) Add NH_4Cl to remaining bottle
  - (4) Open remaining bottle after 3 days
    - a. Analyze for general DBPs
    - b. Analyze for carbonyl compounds
    - c. Analyze by BOM assays (only 2 of 8 runs)

Byproduct and Organic Compound Analyses

General DBP analysis (about 250 mL required) as follows:

Sample no.	Analysis	Quench	Volume
1	Oxidant residual	None	100 mL
2	UV absorbance (UQ)	None	10 mL
3	THM	Na_2AsO_2	40 mL
4	HAA	Na_2AsO_2	60 mL
5	TOC	Na_2AsO_2	20 mL
6	UV absorbance (Q)	Na_2AsO_2	10 mL

Carbonyl Compounds

Sample no.	Analysis	Quench	Volume
1	Aldehydes	Na_2AsO_2	60 mL
2	Ketoacids	Na_2AsO_2	60 mL

BOM Assays

Sample no.	Analysis	Quench	Volume
1	BDOC	Sodium sulfite	100 mL
2	AOC	Sodium sulfite	100 mL

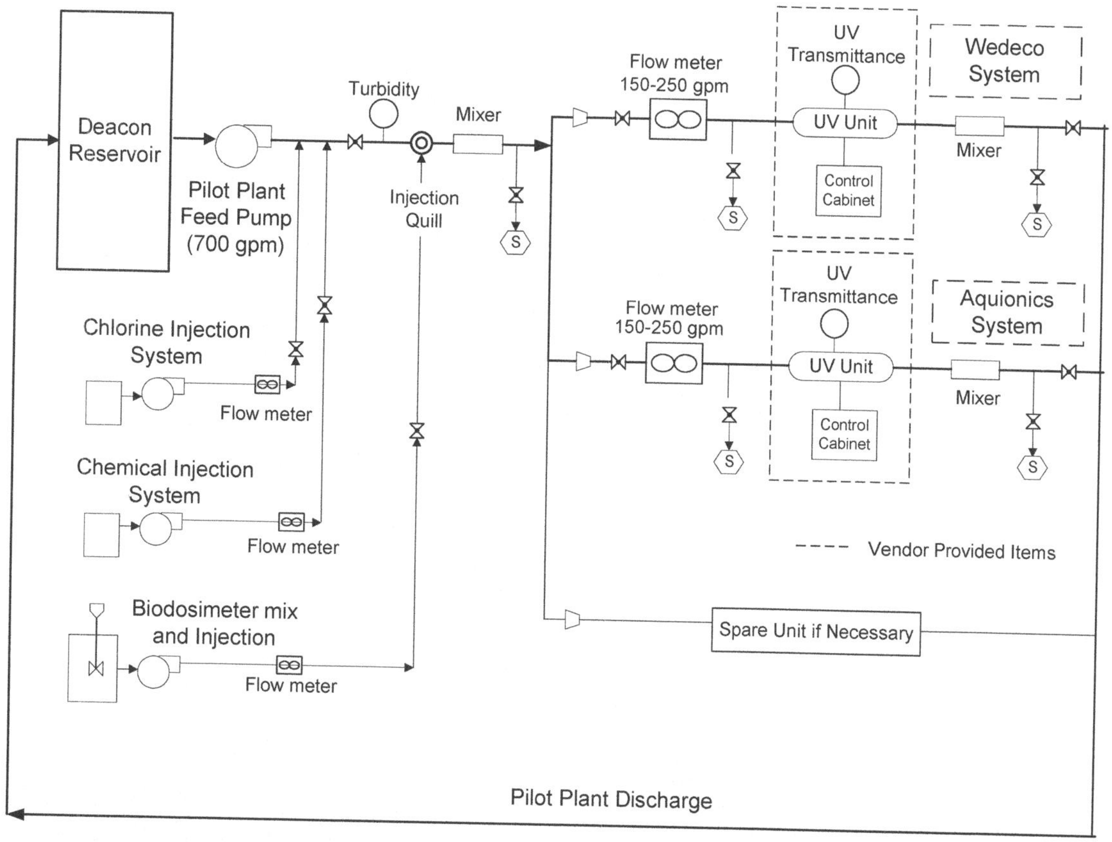

Figure 3.1 Pilot plant process flow schematic

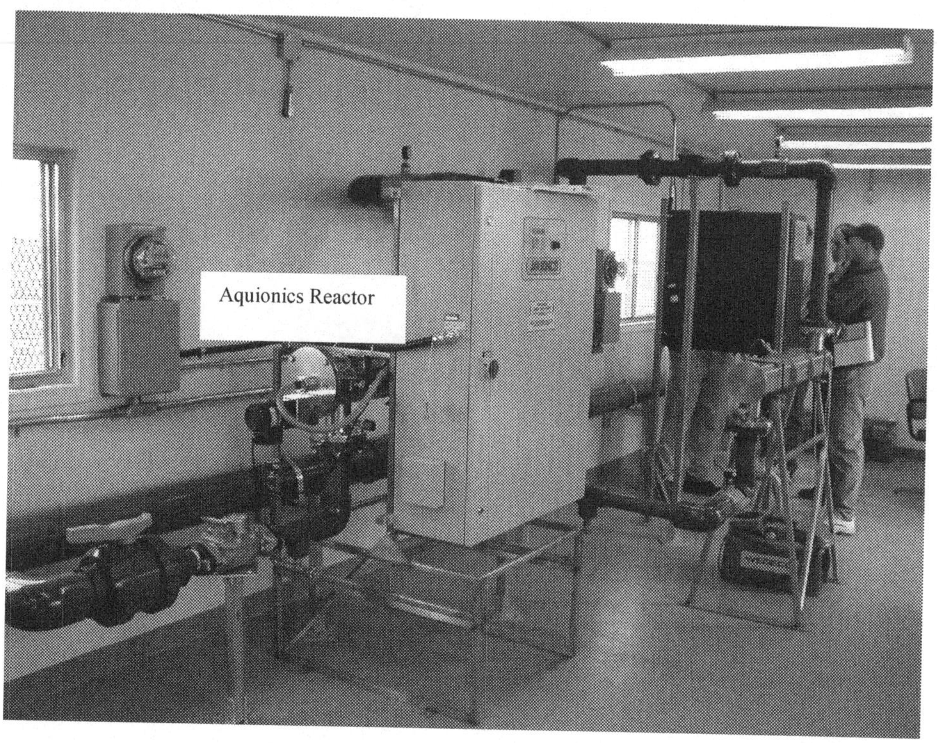

Figure 3.2 Aquionics UV pilot reactor

Figure 3.3 Wedeco-Ideal Horizons UV pilot reactor

Figure 3.4 Calibration and cleaning of UV units

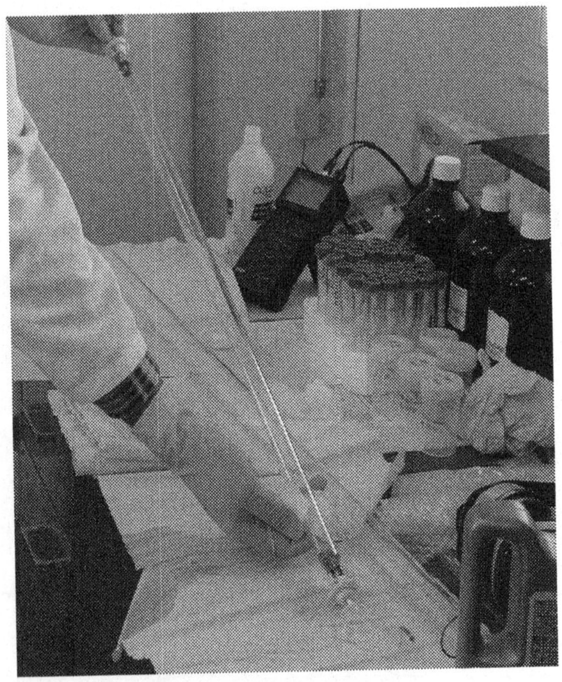

Figure 3.5 Cleaning of UV lamp

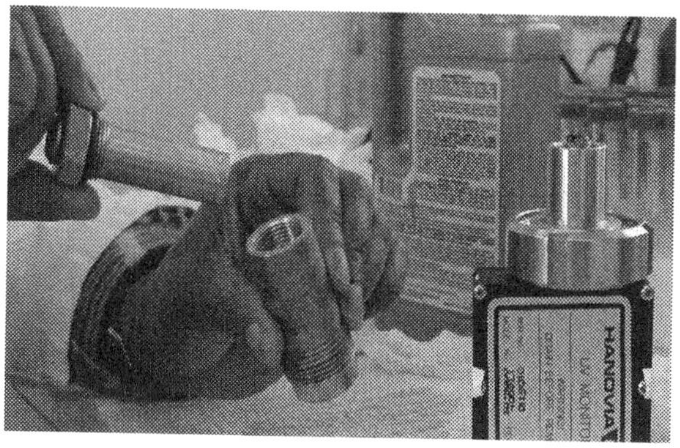

Figure 3.6 Checking UV sensor and monitor

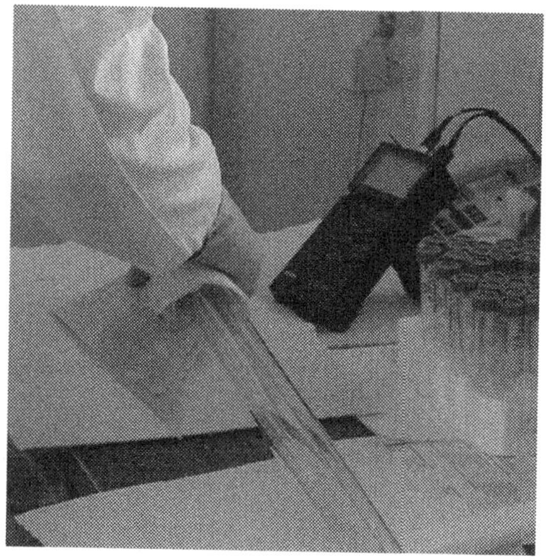

Figure 3.7 Cleaning UV quartz sleeve

Figure 3.8 Collecting sediment for turbidity testing from Deacon Reservoir during the winter with a diver under the ice

Figure 3.9 Sediment being collected for turbidity spiking

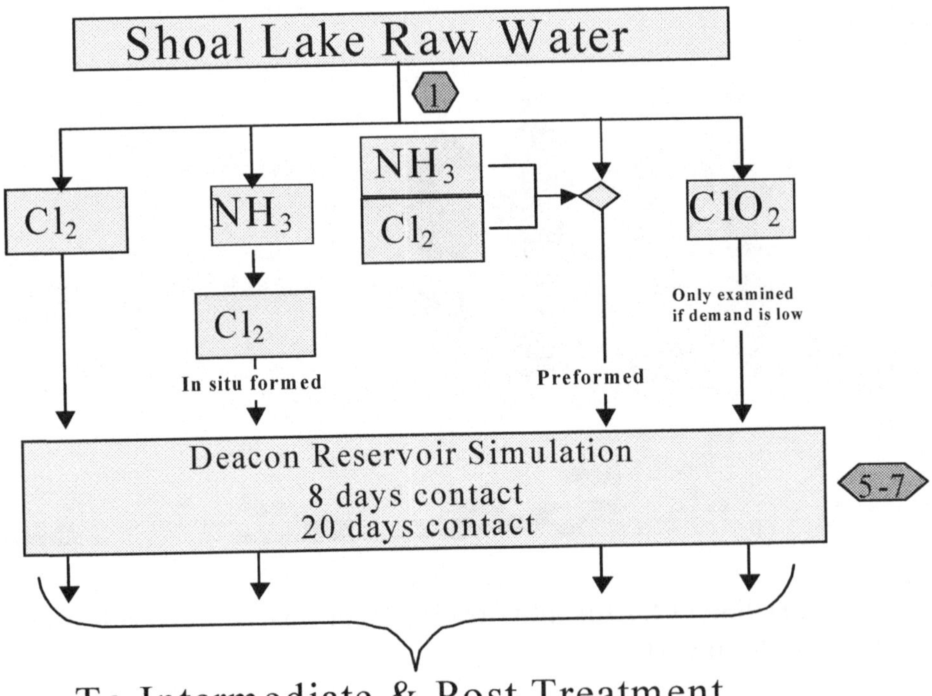

Figure 3.10 Pretreatment bench-scale experimental schematic

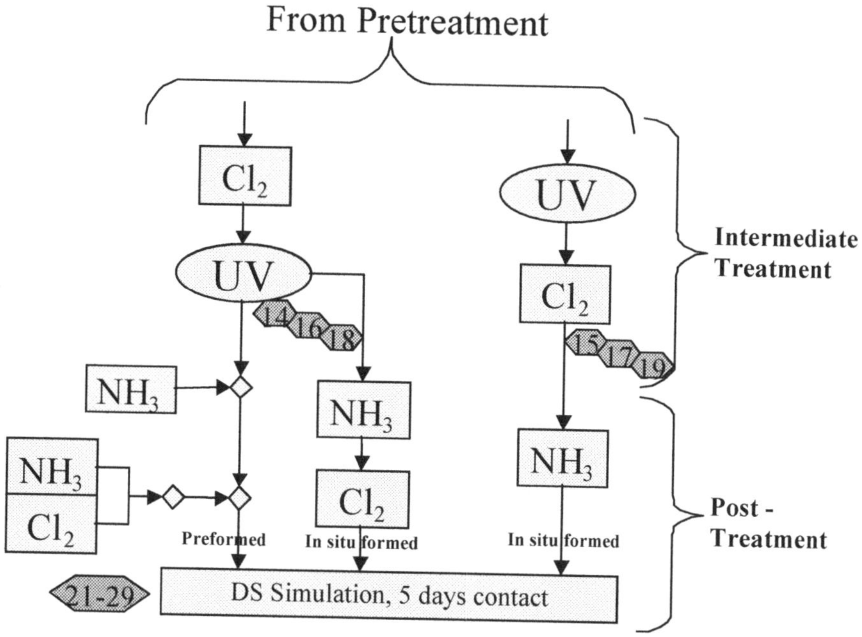

Figure 3.11 Post-treatment bench-scale experimental schematic

CHAPTER 4
UV DISINFECTION RESULTS AND ANALYSIS

SUMMARY

The specific goals and objectives of pilot testing and inactivation studies were as follows:

- Determine the impact of algae and TOC on lamp fouling characteristics and cleaning needs over a 1-year period
- Determine the impact of turbidity on UV disinfection efficacy
- Determine the impact of algae on UV disinfection efficacy
- Determine the impact of TOC on UV disinfection efficacy
- Determine the viability of UV for primary disinfection of Winnipeg's unfiltered water supply

Operational Considerations

The most significant influence on reactor performance observed during this study was the result of cleaning the quartz sleeves. Periodic and relatively frequent cleaning of the sleeves appears to be the most desirable maintenance plan because the sensors seemed to be relatively ineffective in indicating actual reactor performance. From the data, it appears that the Aquionics lamp and Wedeco lamps can operate for approximately 2,000 hours before a cleaning is required. Lamps and sleeves for both systems were changed after approximately 5,000 hours of operation.

Equipment intensity sensors did not track well during the study. Sensors can be a valuable tool in monitoring system performance; however, from the challenges, it is also clear that the sensors can fail to describe what is happening in the system. The unreliable sensor results were due to the pilot-scale equipment and a series of operational difficulties encountered by both UV reactors during the pilot testing. Regardless, sensor performance bears watching in full-scale UV system selection, design, and operation. It is essential during UV reactor validation and long-term operation that the sensors meet tight specifications initially and are proven periodically to be calibrated and responsive to significant changes in performance (RED) caused by such common factors as lamp fouling, flow changes, lamp aging, etc.

Inactivation Studies

The biodosimetry results showed that the UV systems performed predictably on the Deacon Reservoir system. Inactivation and dose targets were achieved on ambient water quality conditions, and both reactors delivered the desired dose to the water. The data also showed the importance of routine lamp sleeve cleaning.

The simulated turbidity spiking event in Inactivation Study No. 4 indicated that the sediments from the reservoir had little to no effect on the RED compared to the cleaning data at 2.5 NTU. There was also no effect on the CB study.

For Inactivation Study No. 5, the turbidity spiking indicated no significant change in reactor performance at levels up to 4 NTU. Because Winnipeg's raw water rarely exceeds 2.0 NTU, turbidity should not be an issue with respect to reactor performance and inactivation.

These results also support the use of UV to achieve *Cryptosporidium* oocyst and *Giardia* cyst inactivation on unfiltered drinking-water systems in compliance with USEPA unfiltered regulations.

Increasing TOC resulted in decreasing UVT. This was expected as TOC will affect UV absorbance. This effect on the UV reactors is expected to be minor due to the narrow range of TOC values (9 to 11 mg/L) typical for Winnipeg's water. However, if a source has widely varying TOC, there will be an impact on UV operations and UV reactors will require higher energy during elevated TOC levels in the Deacon Reservoir to maintain dose delivery.

There is a consistent correlation between plankton counts and UVT. UVT values range between 74.6 and 77.1 percent for various plankton levels. This suggests that UV reactors will require higher energy during elevated plankton levels in the Deacon Reservoir to maintain dose delivery. However, once the UV intensity is corrected, the MS-2 has the same response to delivered UV dose as the samples with lower plankton counts. This suggests the plankton is not shielding or otherwise protecting the MS-2 from inactivation by the UV light.

The results from this study confirm that UV is viable for primary disinfection of the City's unfiltered supply regardless of seasonal water quality variations.

INTRODUCTION

This chapter presents the UV inactivation testing results and the long-term UV pilot operation data collected over 14 months at Winnipeg, Manitoba. Five inactivation tests covered a wide range in water quality. Both field and laboratory tests were conducted to determine the impact of raw water conditions (e.g., turbidity, algae, and TOC) on UV disinfection efficacy and the impacts of long-term operations (e.g., lamp aging and sleeve fouling). The results and conclusions are presented in the following sections:

- Inactivation Studies
- Impact of Turbidity on UV Efficacy
- Impact of TOC and Algae on UV Effectiveness
- Operational Considerations
- Conclusions

With respect to lamp cleaning, it is important to note that the automatic lamp wiper feature for the Aquionics unit was turned off so that a lamp fouling rate comparison could be made to the Wedeco unit, which did not have this feature.

INACTIVATION STUDIES

This section presents and discusses the results of the MS-2 bacteriophage inactivation studies (referred to as MS-2 biodosimetry) performed on the Wedeco-Ideal Horizons and Aquionics UV reactors. Five inactivation tests were conducted during the 14-month study period. The City's staff operated the UV reactors and carried out the field portion of the inactivation tests. The UNH team conducted the associated CB experiments and all MS-2 analyses. This section summarizes the five UNH inactivation studies.

The purpose of the MS-2 biodosimetry was to evaluate the UV dose delivered by the Wedeco and Aquionics units over time as a means of demonstrating system performance and to assist in identifying operation and maintenance requirements. The Wedeco UV system uses low-pressure, high-intensity lamps, while the Aquionics system uses medium-pressure lamps. Testing with these different lamps provided information on the two most common types of UV systems currently used for drinking-water disinfection (low pressure high output [LPHO] and medium pressure [MP]).

The five inactivation studies presented below were performed between June 6, 2001, and July 29, 2002. Each inactivation study was divided into five general categories:

- Raw Water Quality
- UV Scans
- CB Studies
- MS-2 Field Challenge Test
- Summary and Conclusions

Inactivation Study No. 1

Inactivation Study No. 1 includes two MS-2 biodosimetry test runs. The first, conducted on June 6, 2001, reflects a MS-2 biodosimetry test completed after 24 hours of operation. The second test, conducted on June 11, 2001, assessed the reactor after 144 hours of operation. The short period between the two tests was to assess difference in UV lamp performance after a typical burn-in period of 100 hours after startup.

Raw Water Quality

Pilot raw water (Deacon Reservoir outlet) quality is presented in Table 4.1. Based on the water temperature, the water is considered to be transitioning from cool (4 to 15 °C) to warm (>15 °C). Turbidity and TOC levels are typical for this period.

Table 4.1
Pilot raw water quality for June 6, 2001, and June 11, 2001

Water quality parameter	Date	
	June 6, 2001	June 11, 2001
Temperature (°C)	14.5	15.8
Turbidity (NTU)	0.70	0.60
pH (units)	8.1	8.1
UVT (%)	76.8	78.4
TOC (mg/L)	9.0	8.0
Alkalinity (mg/L as $CaCO_3$)	74	76
Hardness (mg/L as $CaCO_3$)	86	86
Plankton counts (cells/mL)	7100 (BG: 43%, D: 28%)	3870 (BG: 40%, PF: 38%)

BG – blue-green; D – diatoms; PF – pigmented flagellate

UV Scans

The UV scan results are presented in Figure 4.1. The scans show the typical featureless curves of increasing UV absorbance (decreasing UVT) with decreasing wavelength. UV influent scans showed the water matrix to be nearly identical over the time between biodosimetry runs from June 6, 2001, to June 11, 2001. The featureless pattern is a good indication that there are no unusual UV interferences from inorganics, natural organics, or synthetic organic chemicals (SOCs).

CB Studies

The CB study results are presented in Figure 4.2. The data resulted in a dose-response given by the following equation:

$$\log \text{MS-2 Inactivation} = 0.0494(\text{RED}) + 0.294 \ (R^2 = 0.985)$$

This result is typical of CB studies conducted on natural water sources. Additionally, the zero dose showed no appreciable variance from the linear regression. The R^2 value of 0.985 indicates an excellent linear correlation of the data (1.0 indicating perfect correlation).

Field Challenge Test

From MS-2 challenge testing, pilot log reductions are presented in Table 4.2. The RED values for each condition were determined for each reactor.

Summary and Conclusions – Inactivation Study No. 1

Operating the Aquionics reactor at 240 gpm and the lamp power level at 1, the UV dose determined by MS-2 biodosimetry (RED) was 19.6 mJ/cm^2 after 24 hours and 25.5 mJ/cm^2 after 144 hours. This slight increase in RED is not statistically significant but may reflect improved lamp stability since the first RED measurement was taken before the 100-hour burn-in and the second was taken after the prescribed burn-in period. Because the desired experimental dose for this unit was 40 mJ/cm^2, the lamp power was adjusted so that a RED closer to 40 mJ/cm^2 would be achieved for extended operation and subsequent inactivation tests. The most likely causes for the difference between the measured RED and the estimated dose provided by the manufacturer for the given operating conditions (i.e., an intended dose of 40 mJ/cm^2) are the actual inlet hydraulics, the actual influent UVT, and the actual lamp output.

Table 4.2
Pilot log reductions for Inactivation Study No. 1

Sample identification	June 7, 2001	June 11, 2001
Aquionics log reduction* round 1	0.97	1.34
Aquionics log reduction round 2	0.97	1.20
Aquionics log reduction round 3	0.91	1.07
Aquionics log reduction UV off	0.01	0.03

(continued)

Table 4.2 (continued)

Sample identification	June 7, 2001	June 11, 2001
Wedeco log reduction round 1	2.04	1.88
Wedeco log reduction round 2	1.89	1.64
Wedeco log reduction round 3	1.94	1.68
Wedeco log reduction UV off	-0.01	0.04

* log reduction = log influent – log effluent

Operating the Wedeco reactor at 115 gpm, the UV dose determined by MS-2 biodosimetry was 41.0 mJ/cm^2 after 24 hours and 36.1 mJ/cm^2 after 144 hours. The slight decrease in RED between the two runs is not statistically significant but may reflect greater lamp stability. Because the desired experimental dose for this unit is 40 mJ/cm^2, the flow through the reactor was adjusted to 86 gpm so that a RED closer to 40 mJ/cm^2 would be achieved for extended operation and subsequent inactivation tests. The most likely causes for the difference between the RED and the operating conditions provided by the manufacturer (i.e., an intended dose of 40 mJ/cm^2) are hydraulics, influent UVT, and actual lamp output.

Inactivation Study No. 2

Inactivation Study No. 2 includes two MS-2 biodosimetry test runs. The first run, conducted on August 28, 2001, reflects a MS-2 biodosimetry test completed before cleaning the quartz sleeves in the UV reactors. The lamps were in operation for 1,765 and 1,846 hours for the Aquionics and Wedeco units, respectively. The second biodosimetry test, conducted 24 hours later, assessed the reactors' performance after cleaning the quartz sleeves. CB studies were run on the water samples taken during this time. In addition, water samples from Shoal Lake, which had elevated levels of natural algae, were used in a CB study to observe the effects of algae on UV performance. The UV sensor values for the two units remained relatively constant during the testing, which was unexpected.

Raw Water Quality

Pilot raw water (Deacon Reservoir outlet) quality is presented in Table 4.3. Based on the water temperature, the water is considered to be warm (>15 °C). Turbidity and TOC levels are typical for this period.

UV Scans

The UV scans for Inactivation Study No. 2 are presented in Figure 4.3. Scans for Study No. 1 are shown for comparison. UV influent scans showed the water matrix to be nearly identical over the time between biodosimetry runs from June 6, 2001, to August 29, 2001. The featureless pattern is a good indication that there are no unusual UV interferences from inorganics, natural organics, or synthetic organic compounds. A sample taken from Shoal Lake had elevated levels of plankton, ~70,000 organisms per mL, and showed an increase in UV absorbance. The increase likely resulted from UV absorbance by organic matter, such as exocellular polymeric materials, related to the plankton. This increased UV absorbance is noteworthy, and if the same effect occurred in Deacon Reservoir, then it would need to be accounted for in a full-scale UV equipment design.

Table 4.3
Pilot raw water quality for August 29, 2001

Water quality parameter	August 29, 2001
Temperature (°C)	22
Turbidity (NTU)	0.70
pH (units)	8.1
UVT (%)	79.4
TOC (mg/L)	9.0
Alkalinity (mg/L as $CaCO_3$)	72
Hardness (mg/L as $CaCO_3$)	77
Plankton counts (cells/mL)	18,200

CB Studies

The CB data for Inactivation Study No. 2 are presented in Figure 4.4. The Deacon Reservoir challenge samples and Shoal Lake challenge sample data indicate a typical linear dose-response observed for MS-2 bacteriophage virus. Statistical analysis of these data suggests that none of the curves are significantly different from one another. The slight differences observed for the June 7, 2001, curve reflect typical variations in microbial analyses, as well as slight recalibrations of the UV radiometers between the June 7, 2001, and the August 28, 2001, studies. It is interesting to note that the Shoal Lake sample microorganism response was not significantly affected by the presence of the plankton (18,200 cells/mL). In other words, once the background UV absorbance is corrected, the MS-2 has the same response to delivered UV dose as the samples where there are no high plankton counts. This suggests the plankton is not shielding or otherwise protecting the MS-2 phage from inactivation by the UV. This result is similar to findings presented by several researchers (Malley and Passantino, 2001; Christensen and Linden, 2001).

Field Challenge Tests

Operating the Aquionics reactor at 255–262 gpm and the lamp power level at 2, the UV dose determined by MS-2 biodosimetry (RED) was 28.7 mJ/cm^2 before lamp cleaning and 52.5 mJ/cm^2 after lamp cleaning (see Table 4.4). This confirms that the cleaning was effective and that the unit is delivering a dose in the desired range of 40 mJ/cm^2. It also demonstrates the importance of cleaning to maintain UV system performance.

Table 4.4
Pilot log reductions for Inactivation Test Nos. 1 and 2

	June 7, 2001	June 11, 2001	August 28, 2001 Pre-clean	August 29, 2001 Post-clean
Aquionics log reduction replicate 1	0.97	1.34	1.19	2.00
Aquionics log reduction replicate 2	0.97	1.20	1.31	2.13
Aquionics log reduction replicate 3	0.91	1.07	0.95	2.24
Geometric mean log reduction	0.95	1.20	1.14	2.12
Reduction equivalent dose (mJ/cm^2)	19.6	25.5	28.7	52.5
Control: Aquionics log reduction UV off	0.01	0.03	-0.15	0.12
Wedeco log reduction replicate 1	2.04	1.88	1.77	3.27
Wedeco log reduction replicate 2	1.89	1.64	1.48	3.10
Wedeco log reduction replicate 3	1.94	1.68	1.86	3.20
Geometric mean log reduction	1.96	1.73	1.70	3.19
Reduction equivalent dose (mJ/cm^2)	41.0	36.1	42.8	78.9
Control: Wedeco log reduction UV off	-0.01	0.04	0.06	-0.15

Operating the Wedeco reactor at 85 to 86 gpm, the UV dose determined by MS-2 biodosimetry (RED) was 42.8 mJ/cm^2 before lamp cleaning and 78.9 mJ/cm^2 after lamp cleaning (see Table 4.4). Because the Wedeco unit has only one power level, the unit is meant to operate at a higher dose at the beginning of a cleaning cycle so that the minimum dose of 40 mJ/cm^2 is achieved when the lamps need cleaning. As with the Aquionics unit, the data demonstrate the importance of cleaning to maintain the UV system performance. The Wedeco and the Aquionics systems differ in that Aquionics has a wiper and Wedeco relies on manual chemical cleaning with the system offline. However, because the wiper on the Aquionics unit was parked and not used during this testing, the two systems had similar fouling characteristics and need for cleaning.

Summary and Conclusions – Inactivation Study No. 2

The biodosimetry results demonstrate that the UV systems are performing predictably on Deacon Reservoir water. The data reinforce the importance of routine lamp sleeve cleaning. The batch bench-scale CB data for the high-plankton sample taken from Shoal Lake suggest that high plankton counts may not be a problem as long as the increased UV absorbance of the water is accounted for in the delivered UV dose.

Inactivation Study No. 3

Inactivation Study No. 3 includes two MS-2 biodosimetry runs, which occurred on October 31, 2001. The first biodosimetry test, referred to as 10/31/01A, was performed before cleaning the quartz tubes in the UV reactors. The second biodosimetry test, 10/31/01B, assessed the reactors' performance after 4 additional hours of operation and after cleaning the quartz sleeves. Prior to test A, the lamps had been operating for 3,100 and 3,186 hours, for the Aquionics and Wedeco units, respectively. A CB study was conducted on a water sample taken during this same time.

Raw Water Quality

Pilot raw water (Deacon Reservoir outlet) quality is presented in Table 4.5. Based on the water temperature, the water is considered to be cold (<4 °C). Turbidity and TOC levels are typical for this period.

Table 4.5
Pilot raw water quality for October 29, 2001

Water quality parameter	October 29, 2001
Temperature (°C)	3.3
Turbidity (NTU)	1.30
pH (units)	8.4
UVT (%)	75.7
TOC (mg/L)	10
Alkalinity (mg/L as $CaCO_3$)	73
Hardness (mg/L as $CaCO_3$)	79
Plankton counts (cells/mL)	27,160

UV Scans

UV scans showed the water matrix to be nearly identical over the time between biodosimetry runs from June 6, 2001, to October 31, 2001. (See Figure 4.5, previous scans are shown for comparison.) The featureless pattern is a good indication that there are no unusual UV interferences from inorganics, natural organics, or synthetic organic compounds. It should be noted that the Challenge No. 3 sample did reflect the greater absorbances due to the higher turbidity of 1.30 NTU and plankton count of 27,160 cells/mL as compared to Challenge No. 2, where turbidity was 0.70 NTU and plankton was 18,200 cells/mL.

CB Studies

The CB challenge on the October 31, 2001, water indicated the typical linear dose-response observed for MS-2 bacteriophage virus (see Figure 4.6). The concurrent trials on deionized water also showed no atypical results.

Field Challenge Tests

There was a significant difference in performance between Challenge No. 2 and Challenge No. 3, as seen in Table 4.6 and depicted graphically in Figure 4.7. Operationally, power and flow adjustments were made between Challenge No. 1 and Challenge No. 2; there were no adjustments made to power or flow between Challenge No. 2 and Challenge No. 3.

Table 4.6
Pilot log reductions for Aquionics and Wedeco UV pilot units, Test Runs 1, 2, and 3

	Challenge No. 1		Challenge No. 2		Challenge No. 3	
	6/7/01 24 hrs. op.	6/11/01 144 hrs. op.	8/28/01 Pre-clean	8/29/01 Post-clean	10/31/01A Pre-clean	10/31/01B Post-clean
Aquionics log reduction replicate 1	0.97	1.34	1.19	2.00	1.06	1.43
Aquionics log reduction replicate 2	0.97	1.20	1.31	2.13	1.05	1.40
Aquionics log reduction replicate 3	0.91	1.07	0.95	2.24	1.10	1.26
Geometric mean log reduction	0.95	1.20	1.14	2.12	1.07	1.36
Reduction equivalent dose (mJ/cm^2)	19.6	25.5	28.7	52.5	25.4	32.4
Control: Aquionics log reduction UV off	0.01	0.03	-0.15	0.12	0.19	-0.12
UVT (%)		78.4		79.4		75.7
Wedeco log reduction replicate 1	2.04	1.88	1.77	3.27	1.40	1.54
Wedeco log reduction replicate 2	1.89	1.64	1.48	3.10	0.65	1.56
Wedeco log reduction replicate 3	1.94	1.68	1.86	3.20	1.44	1.98
Geometric mean log reduction	1.96	1.73	1.70	3.19	1.09	1.69
Reduction equivalent dose (mJ/cm2)	41.0	36.1	42.8	78.9	32.3	40.0
Control: Wedeco log reduction UV off	-0.01	0.04	0.06	-0.15	0.31	0.28

Several conclusions can be derived from Figure 4.7. While operating the Aquionics reactor at 260 gpm and setting the lamp power level at 2 (Challenge No. 3), the RED was 25.4 mJ/cm^2 before lamp cleaning and 32.4 mJ/cm^2 after lamp cleaning. This shows that the cleaning was effective in increasing the RED. However, the system failed to reach the target RED of 40 mJ/cm^2. The power level was increased to the Aquionics unit to attain the desired dose level for subsequent operations and testing. Operating the Wedeco reactor at 85 gpm, the RED was 32.3 mJ/cm^2 before lamp cleaning and 40.0 mJ/cm^2 after lamp cleaning. This demonstrates that the cleaning was effective in increasing the RED; however, this system failed to reach the target RED of 40 mJ/cm^2 for the entire period between cleanings.

Summary and Conclusions – Inactivation Study No. 3

Challenge Study No. 3's biodosimetry results indicate that the UV system's inactivation performance decreased over time. There are several potential reasons for this drop in performance. The UV lamps may have lost intensity. In the case of the Aquionics unit, the intensity sensor did not indicate a decrease in intensity; but the sensor also failed to indicate an increase in intensity after the cleaning in Challenge No. 2 or Challenge No. 3, yet the reactor's performance clearly increased after the cleanings. It is possible that this sensor failed some time after the power increase (i.e., post-Challenge No. 1). The Wedeco intensity sensor measured a decrease in lamp performance over time, but it may have failed before Challenge No. 3 because it did not indicate an increase in intensity after the cleaning as part of Challenge No. 3. The negative sensor response was in disagreement with the reactor's clear improvement in disinfection efficiency after the cleanings in Challenge No. 2 and Challenge No. 3.

Another possible reason for some or all of the decrease in performance may be fouling of the sleeves. If the sleeves are fouling and not getting fully cleaned, they could be causing a decrease in UV intensity and a subsequent decrease in inactivation. Another factor that can account for some of the observed decrease in performance is due to a change in the water matrix since the UVT drops to 76 percent in Challenge No. 3. This decrease in UVT of about 4 percent in the full-scale reactors can have a significant effect on performance.

Inactivation Study No. 4

For Inactivation Study No. 4, three MS-2 biodosimetry runs were performed from March 12, 2002, to March 13, 2002. The first biodosimetry, referred to as the 3/12/02A trial, was performed before the cleaning of the quartz sleeves in the UV reactors. New lamps were installed after Inactivation Study No. 3 just before the start of Inactivation Study No. 4. The tests were performed on lamps with 1,971 and 1,985 hours of operation for the Aquionics and Wedeco units, respectively. The second biodosimetry, 3/13/02B, reflects the reactors' performance after cleaning the quartz sleeves and 21 more hours of operation. The third biodosimetry, 3/13/02C, reflects reactor performance postcleaning with an influent spike of natural sediments to achieve a turbidity of 2.5 NTU. Three CB studies were performed on the water samples taken during this time: two studies on water without the turbidity spike and one with the 2.5 NTU water.

Raw Water Quality

Pilot raw water quality is presented in Table 4.7. Based on the water temperature, the water is considered to be cold (<4 °C). Turbidity and TOC levels are typical for this period.

UV Scans

The UV scans for Inactivation Study No. 4 are presented in Figure 4.8. Previous scans are shown for comparison. The scans show typical featureless curves of increasing UV absorbance (decreasing UVT) with decreasing wavelength. UV influent scans showed the water matrix to be nearly identical over the time between biodosimetry runs from June 6, 2001, to March 13, 2002. The scan of the artificially increased turbid water (2.5 NTU) resulted in higher absorbances as expected.

Table 4.7
Pilot raw water quality for March 11, 2002

Water quality parameter	March 11, 2002 Before turbidity spike	After turbidity spike
Temperature (°C)	1.5	1.5
Turbidity (NTU)	0.60	2.5
pH (units)	7.9	7.9
UVT (%)	72.9	72.9
TOC (mg/L)	11	11
Alkalinity (mg/L as $CaCO_3$)	86	86
Hardness (mg/L as $CaCO_3$)	96	96
Plankton counts (cells/mL)	8,300 (PF: 84%, D: 15%)	8,300 (PF: 84%, D: 15%)

BG – blue-green; D – diatoms; PF – pigmented flagellate

CB Studies

The CB data for Inactivation Test No. 4 is presented in Figure 4.9. The CB results indicated typical linear dose-response observed for MS-2 bacteriophage virus. One trend that is apparently developing is the increase in log reduction for the same dose over time. One factor that may account for some of this change may be the recalibration of the laboratory radiometer between Challenge No. 2 and Challenge No. 3.

Field Challenge Tests

Results of the biodosimetry runs are shown in Table 4.8. Figure 4.10 presents the RED results for this and the previous testing.

Table 4.8
Pilot log reductions and RED for Aquionics and Wedeco units

	Challenge No. 4 3/12/02A Pre-clean	 3/13/02B Post-clean	 3/13/02C 2.5 NTU
Aquionics log reduction no. 1	1.45	2.15	2.33
Aquionics log reduction no. 2	1.53	2.16	2.20
Aquionics log reduction no. 3	1.43	1.93	2.27
Geomean log reduction	1.47	2.08	2.27
RED	37.9	56.8	54.0
Log reduction UV off	-0.02	0.00	0.12
Wedeco log reduction no. 1	1.22	1.98	2.05
Wedeco log reduction no. 2	1.30	1.96	1.89
Wedeco log reduction no. 3	1.12	2.04	1.82
Geomean log reduction	1.21	1.99	1.92
RED	31.0	54.9	45.2
Log reduction UV off	0.06	0.10	0.08

The RED values for each condition were determined by using the log reduction values (Table 4.8) and applying those values to the corresponding dose-response curves.

Operating the Aquionics reactor at 260 gpm and setting the lamp power level at 2, the RED was 37.9 mJ/cm^2 before lamp cleaning, 56.8 mJ/cm^2 after lamp cleaning, and 54.0 mJ/cm^2 with the 2.5 NTU water. These data show that for the third straight challenge, cleaning was effective in increasing the RED. (Note that between Challenge No. 1 and Challenge No. 2, the power level was increased from 1 to 2 and new lamps were installed between Challenge No. 3 and Challenge No. 4.) The turbidity showed little influence on the RED.

Operating the Wedeco reactor at 85 gpm, the RED was 31.0 mJ/cm^2 before lamp cleaning, 54.9 mJ/cm^2 after lamp cleaning, and 45.2 mJ/cm^2 with the 2.5 NTU water. These data show that for the third straight challenge, cleaning was effective in increasing the RED. (Note that the reactor flow was reduced from 115 gpm for Challenge No. 1 to 85 gpm for the remaining challenges, and new lamps were installed between Challenge No. 3 and Challenge No. 4.) The turbidity showed little influence on the RED and the reactor was below the target RED of 40 mJ/cm^2 prior to lamp cleaning.

Summary and Conclusions – Inactivation Study No. 4

Challenge No. 4's biodosimetry results indicate that the Aquionics system increased in performance. The most likely cause for this increase in performance would be the installation of new lamps.

Challenge No. 4's biodosimetry results indicate that the Wedeco system has maintained or slightly increased its performance. The lack of a significant increase in performance indicates that the lamps were still performing satisfactorily. The new intensity sensor tracked changes in intensity more accurately (i.e., there was an increase in the intensity reading that paralleled the observed increase in RED).

The turbidity spiking event indicated that the sediments from the reservoir had little to no effect on the RED compared to the cleaning data. There was also no effect observed during the CB study.

It is interesting to compare and contrast the results from Challenge No. 4 with those from Challenge No. 2 since the lamp ages were similar even though the lamps were different. The Aquionics results from the two challenges are similar although the RED values for Challenge No. 4 are a little higher at 56.8 and 54.0 mJ/cm^2 versus 52.5 mJ/cm^2 after cleaning. This slight increase is most likely the result of better reactor control of delivered dose since the sensors were inspected and improved after Challenge No. 3. The Wedeco results found between Challenge No. 2 and Challenge No. 4 show a significant difference after cleaning with an RED of 78.9 mJ/cm^2 in Challenge No. 2 and an RED of 56.8 mJ/cm^2 after Challenge No. 4. The lower results in Challenge No. 4 may be related to the lamp-to-lamp variability that this Wedeco pilot units had been experiencing since there is only one sensor reading for these four closely packed together lamps. This variability leads to variations in sensor output and can lead to variations in the delivered UV dose occurring in the reactor. Looking at all Challenge runs for Wedeco, the value of 78.9 mJ/cm^2 measured for the Wedeco reactor after cleaning appears to be the greatest outlier. No specific explanation for this can be found after close examination of the data.

Inactivation Study No. 5

Inactivation Study No. 5 includes five MS-2 biodosimetry runs performed from July 29, 2002, through July 30, 2002. The first biodosimetry test, referred to as the 7/29/02A trial, was performed before cleaning the quartz sleeves in the UV reactors. The new lamps were in operation for 1,448 and 1,457 hours for the Aquionics and Wedeco units, respectively. The second biodosimetry, referred to as the 7/29/02B trial, assessed the reactors' performance after cleaning the quartz sleeves. The third biodosimetry, referred to as the 7/30/02C trial, assessed the reactors' performance subsequent to cleaning with an influent spike of natural sediments to achieve a target turbidity of 5 NTU (actually turbidity values were about 4 NTU). The fourth biodosimetry, referred to as the 7/30/02D trial, assessed the reactors' performance subsequent to cleaning with an influent spike of natural sediments to achieve a turbidity of 2.5 NTU. A fifth biodosimetry, referred to as the 7/30/02E trial, was performed at 13 NTU to look at very high turbidity spikes; only one sample was collected.

Five CB studies were preformed on water samples taken during this time: one on Deacon Reservoir water without the turbidity spike; three at turbidities of 2.5, 4.0, and 13 NTU; and a fifth performed on deionized water as a control.

Raw Water Quality

Pilot raw water (Deacon Reservoir outlet) quality is presented in Table 4.9. Based on the water temperature, the water is considered to be warm (>15 °C). Turbidity and TOC levels are typical for this period.

UV Scans

The UV scans for Inactivation Test No. 5 are presented in Figure 4.11. Previous scans are shown for comparison. The compilation of scans for the Deacon Reservoir water shows typical featureless curves of increasing UV absorbance (decreasing UVT) with decreasing wavelength. UV influent scans showed the water matrix had measurable fluctuations in UV absorbance at 253.7 nm from June 6, 2001, to July 30, 2002.

Figure 4.12 compares UV scans of the three spiked turbidity tests to UV scans for unspiked Deacon Reservoir water, which was used as the base water for Inactivation Test No. 5. The scans of the artificially increased turbidity waters show higher absorbances, as expected.

Table 4.9
Pilot raw water quality for July 29, 2002

Water quality parameter	July 29, 2002	
	Before turbidity spikes	After turbidity spikes
Temperature (°C)	23.3	23.3
Turbidity (NTU)	1.2	2.5, 4.0, 13.0
pH (units)	7.9	7.9
UVT (%)	77.8	77.8
TOC (mg/L)	9	9
Alkalinity (mg/L as $CaCO_3$)	68	68
Hardness (mg/L as $CaCO_3$)	71	71
Plankton counts (cells/mL)	14600 (PF: 36%, D: 45%)	14600 (PF: 36%, D: 45%)

BG – blue-green; D – diatoms; PF – pigmented flagellate

CB Studies

The CB studies presented in Figure 4.13 indicate typical linear dose-response observed for MS-2 bacteriophage virus. The concurrent trials on deionized water also showed no atypical results.

Field Challenge Tests

Table 4.10 summarizes the RED values for Inactivation Studies No. 1 through 5.

The RED values are presented in Figure 4.14. The data indicate that overall the reactors have been performing better than initially reported. All confidence intervals are reported at 95 percent.

Note that on Figure 4.14, for the Aquionics unit between Challenge No. 1 and Challenge No. 2, the power level was increased from 1 to 2 and new lamps were installed between Challenge No. 3 and Challenge No. 4.

Table 4.10
Comparison of REDs for Aquionics and Wedeco units for Inactivation Study Nos. 1 through 5

	Challenge No. 1		Challenge No.2		Challenge No. 3	
	6/7/01 24 hrs Op	6/11/01 144hrs Op	8/28/01 Pre-clean	8/29/01 Post-clean	10/31/01A Pre-clean	10/31/01B Post-clean
Aquionics RED (mJ/cm^2)	19.6	25.5	28.7	52.5	25.4	32.4
Wedeco RED (mJ/cm^2)	41.0	36.1	42.8	78.9	32.3	40.0

	Challenge No. 4			Challenge No. 5			
	3/12/2002 Pre-clean	3/13/2002 Post-clean	3/13/2002 4.0 NTU	7/29/2002 Post-clean	7/30/2002 2.5 NTU	7/30/2002 4.0 NTU	7/30/2002 13 NTU
Aquionics RED (mJ/cm^2)	37.9	56.8	54.0	36.7	36.7	32.2	26.5
Wedeco RED (mJ/cm^2)	31.0	54.9	45.2	40.4	42.9	39.2	33.2

For the Wedeco unit, the reactor flow was reduced from 115 gpm for Challenge No. 1 to 85 gpm for the remaining challenges and new lamps were installed between Challenge No. 3 and Challenge No. 4.

During this final challenge (No. 5), the Aquionics reactor was operated at the customary flow of 260 gpm, but the lamp power level was increased from 2 to 3. The increased power level had no effect on the RED before cleaning when compared to the fourth challenge performed on March 12 and 13, 2002 (38.7 +/- 2.1 mJ/cm^2 versus 37.9 +/- 1.7 mJ/cm^2). As shown in Figure 4.14 for the Aquionics unit, the addition of sediments to increase the turbidity resulted in small changes in the resulting RED up to 2.5 NTU with these effects increasing for the 4 NTU and being most pronounced for the 13 NTU case, which was the only level that demonstrated a statistically significant reduction in UV performance when compared to the initial operating condition (Figure 4.14).

During the final challenge, the Wedeco reactor was operated at the customary 85 gpm. The RED before cleaning during Challenge No. 5 when compared to the RED before cleaning during Challenge No. 4 increased from 31.0 +/- 2.6 mJ/cm^2 to 45.8 +/- 1.4 mJ/cm^2, which is likely the result of sensor cleaning, sensor board replacement, and complete lamp replacements made on the unit between challenges. As with the Aquionics unit, this was the first challenge where cleaning the quartz sleeves did not result in an increase in performance. The addition of turbidity indicated that increasing turbidity can reduce RED. However, none of the RED values for the turbid waters were statistically different from the initial operating condition (Figure 4.14). The large error bar for the 13 NTU water is due to the reduced degrees of freedom in this experiment. Only a small number of samples could be taken because it was at the end of the spiking run and there was moderate variance in the duplicate plates. As turbidity increases to these higher levels, the MS-2 assay and the plating, in particular, are subject to greater variability.

QA/QC for the trip blanks and pilot unit operation with lamps off during this final challenge, both indicated no significant log losses in MS-2 concentration.

Summary and Conclusions – Inactivation Study No. 5

Challenge No. 5 lends confidence to earlier results achieved throughout the study. Both reactors are delivering the desired dose to the water. The turbidity spiking for the second straight challenge indicated no significant change in reactor performance at levels up to 4 NTU. This indicates that turbidity levels will have very little effect on reactor performance for Winnipeg's water. Because Winnipeg's raw water rarely exceeds 2.0 NTU, turbidity should not be an issue with respect to reactor performance and inactivation.

The Aquionics sensor was extremely sensitive to turbidity fluctuations while spiking sediment. Adjustments on the sediment pump were reflected in the Aquionics' sensor output. It was observed that slight changes in turbidity would result in a change of the sensor reading. However, the sensor had failed in the past as discussed below (Challenge No. 2, Challenge No. 3, and Challenge No. 4) to indicate any significant difference in performance compared to postcleaning, while the biodosimetry did show a significant change.

The Wedeco sensor only measured gross changes in turbidity (0 to 4 to 13 NTU). The data from the challenge indicate that only gross changes in the turbidity resulted in changes in the delivered dose. This sensor indicates what the biodosimetry results confirm, that changes in turbidity have only a small impact on delivered dose. However, this sensor failed in the past

(Challenge No. 3 and Challenge No. 4) to indicate any significant difference in performance pre- and postcleaning, while the biodosimetry results did show a significant improvement.

Sensors are a necessary tool in monitoring system performance. However, based on this testing, it is clear that the sensors can misinterpret what is occurring in the system. The Aquionics sensors reacted to turbidity much more than the Wedeco sensors. The sensors on both systems failed to respond to lamp cleaning during Challenge No. 5 and were inconsistent in their correlation to RED throughout the study. Clearly the sensors are only a part of a UV reactors dose delivery and control system and must be carefully evaluated and interpreted in concert with the flow the reactor is treating, the UVT of the water, the lamp age and lamp power, and the degree of lamp sleeve and sensor window fouling that is occurring.

IMPACT OF TURBIDITY ON UV DISINFECTION EFFICACY

One of the objectives of the collaborative study was to determine the impacts of raw water turbidity on UV effectiveness. However, ambient water conditions did not allow for naturally high turbidity testing. Figure 4.15 presents Deacon Reservoir turbidity data throughout the pilot program.

For Inactivation Study Nos. 1, 2, and 3, ambient water did not exhibit the turbidity levels desired to determine turbidity impacts on UV effectiveness. Consequently, a decision was made to simulate a high-turbidity event by collecting natural sediment from the bottom of the Deacon Reservoir and introducing the sediment into the pilot raw water inlet line. The target turbidity for Inactivation Study No. 4 was 2.5 NTU. Inactivation Study No. 5 further investigated turbidity impacts by simulating three different turbidity levels, 2.5, 4.0, and 13 NTU.

Challenge No. 4 demonstrated that increasing turbidity from approximately 0.6 NTU to 2.5 NTU did not influence the efficiency of either UV disinfection system.

Challenge No. 5 was designed to assess the impact of elevated turbidity on UV disinfection efficacy at 1, 2.5, 4, and 13 NTU. The most important finding was that turbidity as high as 4 NTU appears to have little, if any, influence on UV disinfection efficacy. Interference was not observed until a test was conducted at a turbidity of 13 NTU. These results indicate that unfiltered systems that meet USEPA regulatory requirements for turbidity (i.e., <5 NTU maximum) can employ UV for *Cryptosporidium* oocyst inactivation without having to undergo additional UV reactor validation.

It is important to note that there are uncertainties and limitations when spiking turbidity into a system to simulate a turbidity event and also spiking in the challenge organism. It is unclear the degree to which naturally occurring microorganisms are particle associated in natural high turbidity events at a water treatment plant. Previous studies using MS-2 and particles at UNH have shown through scanning electron microscopy (SEM) that the MS-2 is found to be in the mixture as a free colloid or surface associated. If the MS-2 biodosimeter was completely absorbed and surrounded by an amorphous natural particle and not just free or surface associated, then the results of the biodosimetry testing could be very different (i.e., showing much lower RED values). The degree to which the latter scenario is realistic is beyond the scope of this study and could not be determined. It should be noted, however, that many high turbidity events involve discrete well-defined shaped particles, such as inorganic mineral particles or sediments, and these do not have the capacity to totally engulf the MS-2 biodosimeter. Further, it is not the intent nor within the scope of this study to imply that UV disinfection is an adequate replacement for proper filtration. It is the general feeling of this research team that multiple treatment barriers are essential to adequate public health protection. The decision to filter a given water supply needs to be based on many other carefully considered technical, economic, and political realities, which are far beyond the scope of this research project.

IMPACT OF TOC AND ALGAE ON UV DISINFECTION EFFECTIVENESS

In addition to impacts of turbidity on UV effectiveness, the impacts of TOC and algae are also of particular interest. It is hypothesized that the nature and quantity of TOC will affect UVT and that algae can potentially shield and have a tendency to agglomerate with pathogens, resulting in a reduction in UV efficiency.

Total Organic Carbon

The effects of medium- to high-TOC waters on UV have not been investigated in great detail. Shoal Lake water has moderate TOC levels ranging from 5 to 17 mg/L. While the TOC levels are moderate, the color is typically low, averaging less than 5 color units (CU).

Figure 4.16 presents Deacon outlet (pilot raw water) TOC data throughout the pilot program. The TOC values ranged from 8 mg/L in June 2001 to 12 mg/L in March 2002.

To evaluate the potential effects of TOC on UV effectiveness, two areas of investigation were followed: (1) assess the effects of TOC on UVT and (2) assess the effects of TOC on inactivation. Figure 4.17 presents raw water TOC versus UVT data at various times during the pilot program. In an attempt to isolate TOC effects, dates that have similar turbidity and pH values were selected. For all data points, turbidity ranged between 0.6 and 0.7 NTU, and pH ranged between 7.9 and 8.1. The data show that increasing TOC results in decreasing UVT, as anticipated.

The narrow range of TOC values seen during this study (typically 9 to 11 mg/L) makes it difficult to quantify the effects of TOC on UV inactivation.

Algae Levels

Figure 4.18 presents Deacon outlet (pilot inlet) plankton count data throughout the pilot program. Other than one isolated event in late 2001 (285,000 cells/mL), plankton counts were relatively low (for Deacon Reservoir water) and consistent throughout the study, typically ranging from 3,000 to 30,000 counts/mL. In an attempt to isolate the effects of plankton levels on UVT, weekly water quality data were sorted by raw water samples that had relatively consistent water quality parameters (primarily turbidity and TOC). The resulting data set is presented in Table 4.11.

Figure 4.19 presents the UVT data from Table 4.11. The data suggest that there is a consistent correlation between plankton counts and UVT. UVT values range between 74.6 and 77.1 percent. This suggests that UV reactors will require higher energy during elevated plankton levels.

The data indicate that as the plankton levels increase, the UVT decreases. Because UV dose delivery is partly a function of UVT, plankton count levels will have an effect on UV dose. The information gathered during this study was not adequate to quantify this relationship, as higher plankton counts were not seen in the Deacon Reservoir during the inactivation studies. In general, the higher the plankton counts, the lower the UVT and UV dose.

Table 4.11
UVT versus plankton count data

Date	Turbidity (NTU)	pH	UVT (%/cm)	TOC (mg/L)	Plankton (cells/mL)	Plankton types (%)
Sept 24, 2001	0.9	8.0	76.7	10	28,600	BG (75) PF (14)
Oct 1, 2001	1.2	8.1	74.6	11	285,000	PF (99) BG (01)
Oct 9, 2001	1.3	8.0	75.7	10	46,500	BG (73) D (14)
Oct 15, 2001	1.1	8.0	75.5	10	51,570	PF (50) BG (33)
Oct 22, 2001	1.0	8.1	76.6	10	24,240	PF (48) D (46)
Nov 13, 2001	1.0	8.2	77.1	10	13,800	D (49) PF (40)

BG – blue-green; PF – pigmented flagellates; D – diatoms

During Inactivation Study No. 2, CB tests were conducted on Shoal Lake water and water from the pilot plant. At that time, plankton levels in the Shoal Lake water were approximately 70,000/mL. Figure 4.20 presents the CB results. From Figure 4.20, it is apparent that the Shoal Lake sample microorganism response was not significantly affected by the presence of the plankton. Once the background UV absorbance is corrected, the MS-2 has the same response to delivered UV dose as the samples where there are no high plankton counts. This indicates that the plankton is not physically shielding or otherwise protecting the MS-2 biodosimeter from inactivation by the UV, which is similar to findings presented by several researchers (Malley and Passantino, 2001; Christensen and Linden, 2001).

OPERATIONAL CONSIDERATIONS

The UV pilot plant was operated by the City from June 2001 through September 2002. Operation and maintenance information included the following areas:

- UV intensity sensors
- Lamp cleaning and replacement
- Power consumption

UV Intensity Sensors

UV intensity for the Aquionics and Wedeco units is presented in Figures 4.21 and 4.22. These figures do not indicate typical UV intensity versus time graphs for UV reactors. Several problems occurred during the study, and most of these problems were related to the UV intensity sensors, including sensor board failure, improper sensor calibration, irregular sensor window fouling and poor cleaning, and sensor failure due to UV degradation of the sensing compounds. It was intended that the intensity data would provide lamp cleaning information. However, due to the variability in intensity sensor data, inactivation test results were used to provide operational information on lamp life and lamp cleaning intervals.

Table 4.12 compares the sensor outputs measured during the challenge studies to the RED calculated from the biodosimetry performed at each of those conditions. It is important to recognize that the sensors measure UV intensity (typically in mW/cm^2) and not the delivered UV dose, whereas the RED is a measure of the delivered UV dose (in mJ/cm^2). Therefore, the numerical values of sensor output and RED should never be equivalent. However, a positive correlation between sensor output and RED is expected.

For both the Aquionics and the Wedeco reactors, it could be argued that the sensor output tracks reasonably well with the RED values in most cases because most of the outputs are flat. However, it is rather surprising that in several instances during the challenge studies, neither the Aquionics sensor nor the Wedeco sensor effectively detected the change in RED after cleaning. Whether this result is an artifact of the pilot-scale equipment because many of the lamp and sensor components were experimental or the result occurs because of a series of operational difficulties encountered during the pilot testing, it bears consideration during design and operation of full-scale UV systems. Full-scale UV sensor selection, design, operation, and maintenance should comply with all of the proposed guidelines that will be put forward in the final version of the USEPA UVDGM expected to be released in 2004.

Table 4.12
Sensor outputs for Aquionics and Wedeco units at the end of each field challenge

	Challenge No. 1		Challenge No. 2		Challenge No. 3		Challenge No. 4			Challenge No. 5			
	Pre-burn-in	Post-burn-in	Pre-clean	Post-clean	Pre-clean	Post-clean	Pre-clean	Post-clean	4.0 NTU	Pre-clean	Post-clean	2.5 NTU	4.0 NTU
Aquionics sensor (mW/cm^2)	52	60	64	64	63	64	41	41	36	52	61	57	52
Aquionics RED (mJ/cm^2)	19.6	25.5	28.7	52.5	25.4	32.4	37.9	56.8	54.0	31.1	28.2	27.7	22.8
Wedeco sensor (W/m2)	110	110	67	98	74	75	37	38	34	49	50	47	48
Wedeco RED (mJ/cm^2)	41.0	36.1	42.8	78.9	32.3	40.0	31.0	54.9	45.2	38.3	32.5	33.7	30.0

The sensors also performed quite differently through the turbidity spiking test (Challenge No. 4 and Challenge No. 5). The Aquionics sensor was extremely sensitive to turbidity fluctuations. Adjustments on the sediment pump were reflected in the Aquionics sensor output. Yet, as Table 4.12 indicates, the sensor had failed in the past (Challenge Nos. 2 through 4) to indicate any significant difference in performance pre- and postcleaning, although the biodosimetry showed a significant improvement.

The Wedeco sensor output responded only to gross changes in turbidity (0 to 4 to 13 NTU). The data from the challenge studies indicate that only gross changes in the turbidity resulted in changes in the delivered dose. This sensor indicates what the biodosimetry results confirm, that changes in turbidity have only a small effect on delivered dose. Yet, as Table 4.12 indicates, the sensor had also failed in the past (Challenge No. 3 and Challenge No. 4) to indicate any significant difference in performance pre- and postcleaning, although the biodosimetry results showed a significant improvement.

The data indicate that the sensors were not good indicators of UV dose. Sensors can be a valuable tool in monitoring system performance; however, as indicated in the challenges, it is also clear that the sensors can fail to describe what is happening in the system. The somewhat erratic sensor results presented in Figures 4.21 and 4.22 may be due in part to the pilot-scale experimental equipment and a result of a series of operational difficulties encountered by both UV reactors during the pilot testing. Regardless, sensor performance bears watching in full-scale UV system selection, design, and operation. It is essential during UV reactor validation and long-term operation that the sensors meet tight specifications initially and are proven periodically to be calibrated and responsive to significant changes in performance (RED) caused by such common factors as lamp fouling, flow changes, lamp aging, etc.

Lamp Cleaning and Replacement

The most significant influence on reactor performance observed during this study was the result of cleaning the quartz sleeves on the lamps and properly maintaining the sensors. There were five lamp cleaning cycles for both reactors throughout the pilot program. Table 4.13 summarizes the lamp cleaning cycles and the time between cleanings. While each unit was operated in parallel, lamp operating times between cleanings vary slightly between the two manufacturers due to power outages, system maintenance, and lamp replacements. The data also show the effect of lamp cleaning on the RED values for each reactor.

Table 4.13
Lamp cleaning intervals

	Cleaning 1 (Challenge No. 2)				Cleaning 2 (Challenge No. 3)			
	Clean date	Hrs. of operation	Pre-clean RED	Post-clean RED	Clean date	Hrs. of operation	Pre-clean RED	Post-clean RED
Aquionics	8/29/01	1791	28.7	52.5	10/31/01	3100	25.4	32.4
Wedeco	8/29/01	1872	42.8	78.9	10/31/01	3186	32.3	40.0
	Cleaning 3 (Challenge No. 4)				Cleaning 4 (No challenge)			
	Clean date	Hrs. of operation	Pre-clean RED	Post-clean RED	Clean date	Hrs. of operation	Pre-clean RED	Post-clean RED
Aquionics	3/12/02	1971	37.9	56.8	5/29/02	1698	N/A	N/A
Wedeco	3/12/02	1985	31.0	54.9	5/29/02	1792	N/A	N/A
	Cleaning 5 (Challenge No. 5)							
	Clean date		Hrs. of operation		Pre-clean RED		Post-clean RED	
Aquionics	7/29/02		1448		38.7		36.7	
Wedeco	7/29/02		1457		40.4		42.9	

For the Aquionics unit, the cleaning cycles varied between 1,400 and 3,100 hours. From the data, it appears that the Aquionics lamp can operate for approximately 1,500 hours before a cleaning is required. It should be noted that this cleaning interval is based on the automatic wiper feature being turned off. The initial lamp was replaced on December 20, 2001, after 4,300 hours of operating. The new lamp was operated for 5,300 hours to July 29, 2002.

For the Wedeco unit, the cleaning cycles varied between 1,700 and 3,200 hours. It is important to note that the recommended chemical cleaning procedure for the Wedeco pilot unit of once every 6 months (4,300 hours) was inadequate for cleaning the sleeves. This was evidenced by little change in intensity levels and REDs before and after this type of cleaning procedure. To provide adequate cleaning, the quartz sleeves had to be removed and cleaned by hand using a calcium, lime, and rust (CLR) remover. This required procedure is a concern due to full-scale operations and maintenance (O&M) implications. From the data, it appears that the Wedeco lamps can operate for about 2,000 hours before a cleaning is required in this application. The initial lamp and sleeves were replaced on December 20 after 4,354 hours of operating. The new lamps were operated for 5,249 hours to July 29, 2002.

Power Consumption

From June 5, 2001, through July 29, 2002, the Aquionics unit used 32,187 kW-hr of power in 9,524 hours of operation, or 3.38 kW average power draw for the given conditions of transmittance and to achieve the target dose. This corresponds to 13 W/gpm for average flow of 260 gpm. Converting to kW/mgd, this is 9.02 kW/mgd at an average UVT of 77 percent.

From June 5, 2001, through July 29, 2002, the Wedeco unit used 5,880 kW-hr of power in 9,607 hours of operation, or 0.61 kW average power draw for the given conditions of transmittance and to achieve the target dose. This corresponds to 7.2 W/gpm for an average flow of 85 gpm. Converting to kW/MG gives 5.0 kW/mgd at an average UVT of 77 percent.

This indicates that the low-pressure, high-intensity lamps are more efficient than the medium-pressure lamps if both systems have been validated by bioassay and are providing similar doses. These findings are consistent with the findings of many previous UV studies that compare the two types of lamp technology at full scale. Although power use is a key issue, there are many other tangible and intangible differences between LPHO and MP systems. Careful cost versus benefit comparisons between LPHO and MP systems must be done before deciding which full-scale system to use, and the results will often be site specific.

CONCLUSIONS

Inactivation Studies

The biodosimetry results showed that the UV systems performed predictably on the Deacon Reservoir system. Inactivation and dose targets were achieved on ambient water quality conditions.

The turbidity spiking event simulated in Inactivation Study No. 4 indicated that the sediments from the reservoir had little to no effect on the RED compared to the cleaning data at 2.5 NTU. There was also no effect on the CB study.

For Inactivation Study No. 5, the turbidity spiking indicated no significant change in reactor performance at levels up to 4 NTU. Because Winnipeg's raw water rarely exceeds 2.0 NTU, turbidity should not be an issue with respect to reactor performance and inactivation. These results also support the use of UV to achieve *Cryptosporidium* oocyst and *Giardia* cyst inactivation on unfiltered drinking-water systems in compliance with USEPA unfiltered regulations.

Increasing TOC resulted in decreasing UVT. This is expected as TOC will affect UV absorbance. This effect on the UV reactors is expected to be minor due to the narrow range of TOC values (9 to 11 mg/L) typical for Winnipeg's water. However, if a source has widely varying TOC, there will be an impact on UV operations.

The data suggest that there is a consistent correlation between plankton counts and UVT. UVT values range between 74.6 and 77.1 percent for various plankton levels. This suggests that UV reactors will require higher energy during elevated plankton levels in the Deacon Reservoir to maintain dose delivery. However, it is interesting to note that the Shoal Lake sample microorganism response was not significantly affected by the presence of the plankton. Once the UV intensity is corrected, the MS-2 has the same response to delivered UV dose as the samples where there are no high plankton counts. This suggests the plankton is not shielding or otherwise protecting the MS-2 from inactivation by the UV light.

Operational Considerations

The Aquionics sensor was extremely sensitive to turbidity fluctuations while spiking sediment. Adjustments on the sediment pump were reflected in the Aquionics sensor output. Yet the sensor failed during Challenge No. 2 and Challenge No. 4 to indicate any significant difference in performance pre- and postcleaning, although the biodosimetry showed a significant improvement.

The Wedeco sensor output responded only to gross changes in turbidity (0 to 4 to 13 NTU). The data from the challenge indicate that only gross changes in the turbidity resulted in changes in the delivered dose. This sensor indicates what the biodosimetry results confirm — changes in turbidity have only a small effect on delivered dose. However, the sensor data is not reliable because the sensor failed during Challenge No. 3 and Challenge No. 4 to indicate any significant difference in performance for pre-and postcleaning although biodosimetry results showed a significant improvement.

Equipment intensity sensors did not track well during the study. Sensors can be a valuable tool in monitoring system performance; however, from the challenges, it is also clear that the sensors can fail to describe what is happening in the system. The unreliable sensor results were due to the pilot-scale equipment and a series of operational difficulties encountered by both UV reactors during the pilot testing. Regardless, sensor performance bears watching in full-scale UV system selection, design, and operation. It is essential during UV reactor validation and long-term operation that the sensors meet tight specifications initially and are proven periodically to be calibrated and responsive to significant changes in performance (RED) caused by such common factors as lamp fouling, flow changes, lamp aging, etc.

The most significant influence on reactor performance observed during this study was the result of cleaning the quartz sleeves. Periodic and relatively frequent cleaning of the sleeves appears to be the most desirable maintenance plan because the sensors seemed to be relatively ineffective in indicating actual reactor performance. From the data, it appears that the Aquionics lamp and Wedeco lamps can operate for approximately 1,500 hours before a cleaning is required. Lamps and sleeves for both systems were changed after 4,300 to 4,400 hours of operation.

From June 5, 2001, through July 29, 2002, the Aquionics unit exhibited a 3.38 kW average power draw for given conditions of transmittance to achieve target dose. This corresponds to 9.02 kW/mgd at an average UVT of 77 percent. From June 5, 2001, through July 29, 2002, the Wedeco unit exhibited a 0.61 kW average power draw for given conditions of transmittance to achieve target dose. This corresponds to 5.0 kW/mgd at an average UVT of 77 percent. This indicates that the low-pressure, high-intensity lamps are more efficient than the medium-pressure lamps if both systems have been validated by bioassay and are providing similar doses. These findings are consistent with the findings of many previous UV studies that compare the two types of lamp technology at full scale. Although power use is a key issue, there are many other tangible and intangible differences between LPHO and MP systems. Careful cost versus benefit comparisons between LPHO and MP systems must be done before deciding which full-scale system to use, and the results will often be site specific.

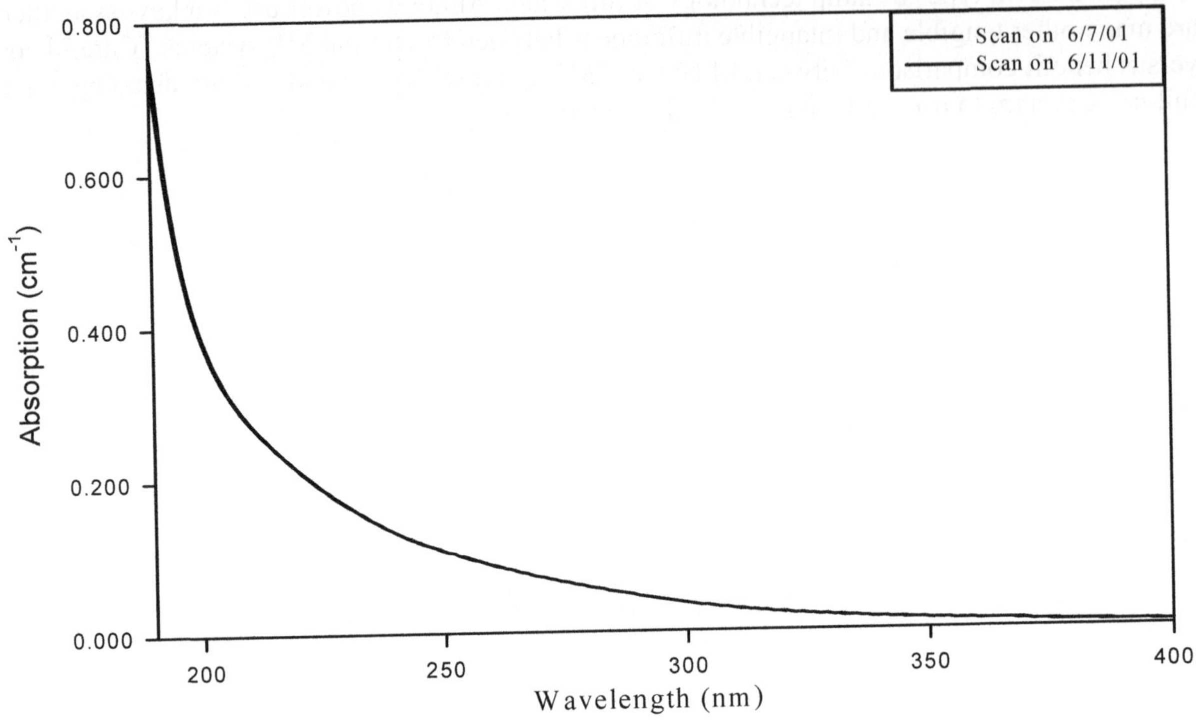

Figure 4.1 UV scans for Inactivation Study No. 1

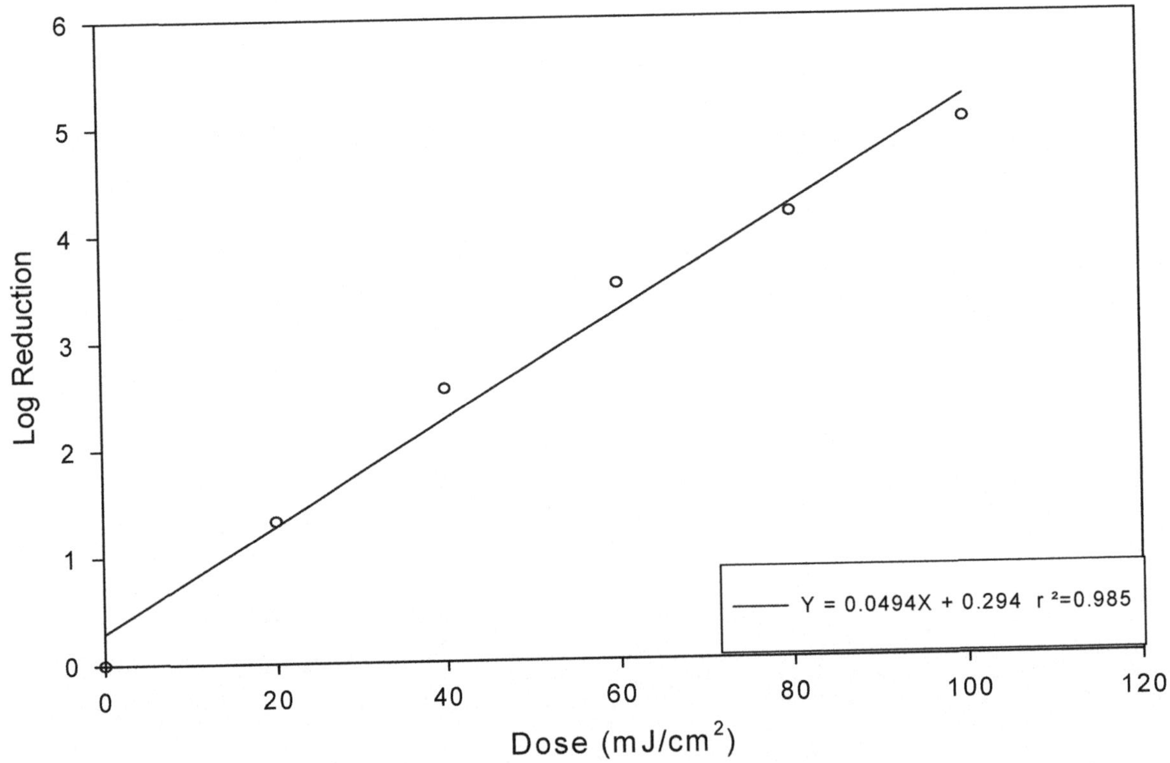

Figure 4.2 Collimated beam results for Inactivation Study No. 1 (June 11, 2001)

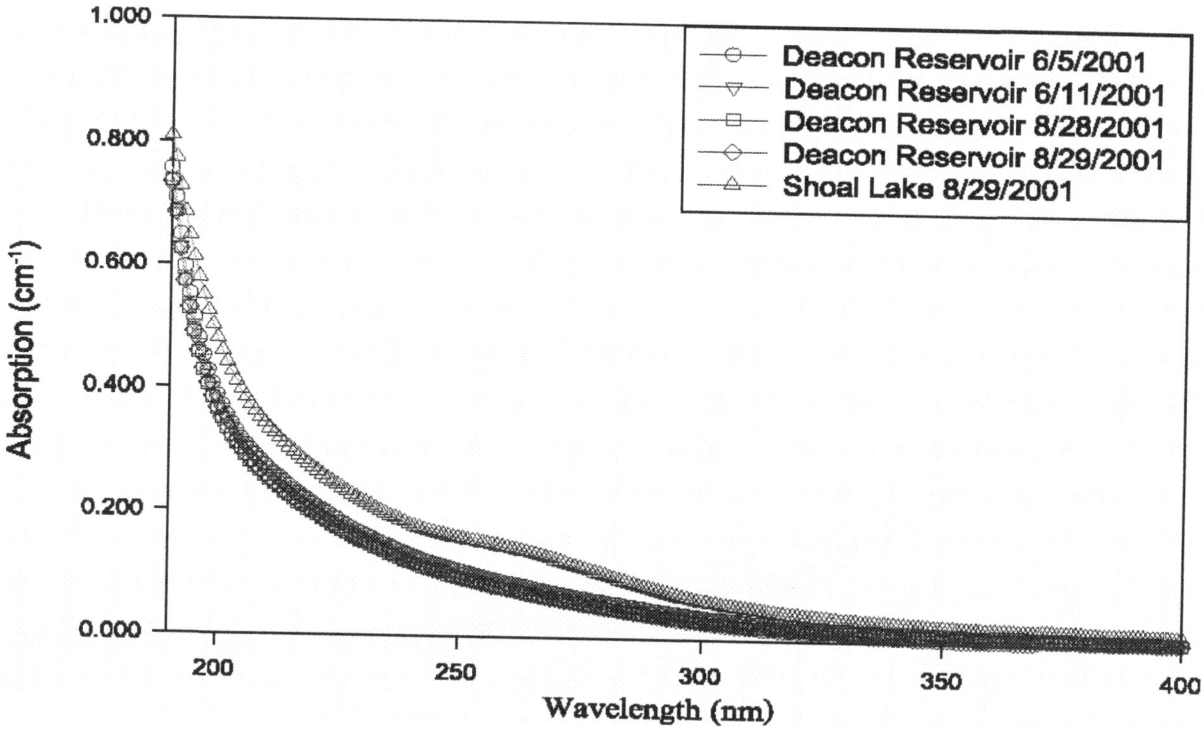

Figure 4.3 UV scans for Inactivation Study Nos. 1 and 2

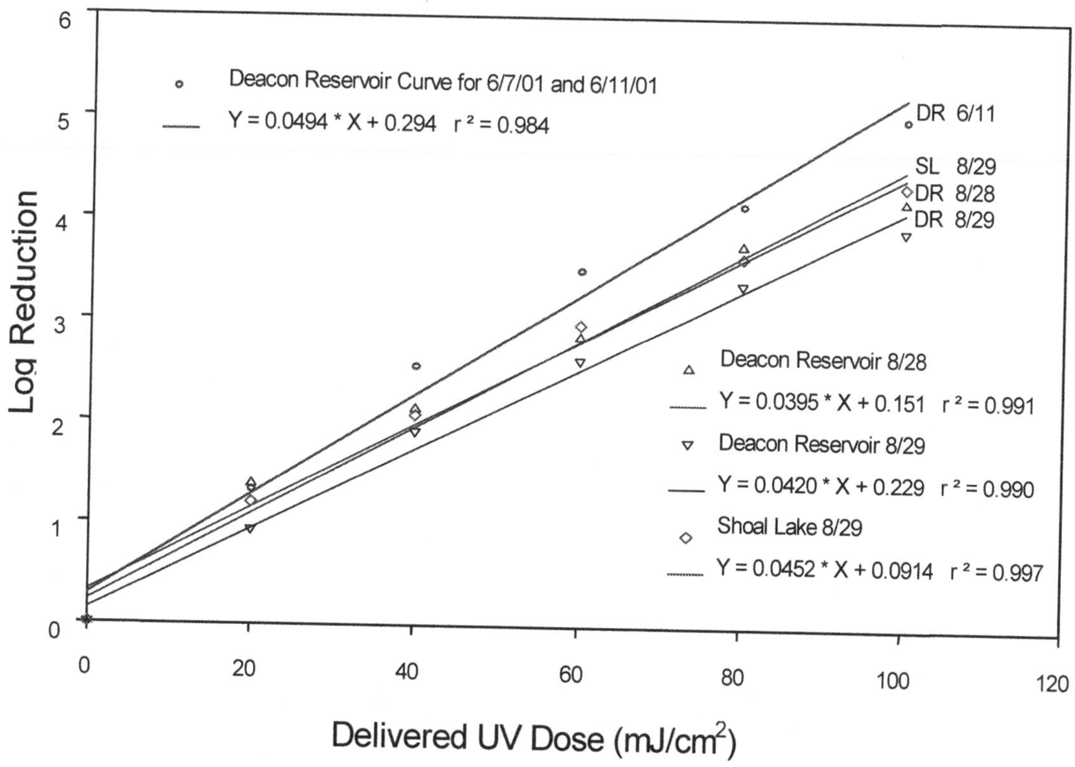

Figure 4.4 Collimated beam results for Inactivation Study Nos. 1 and 2

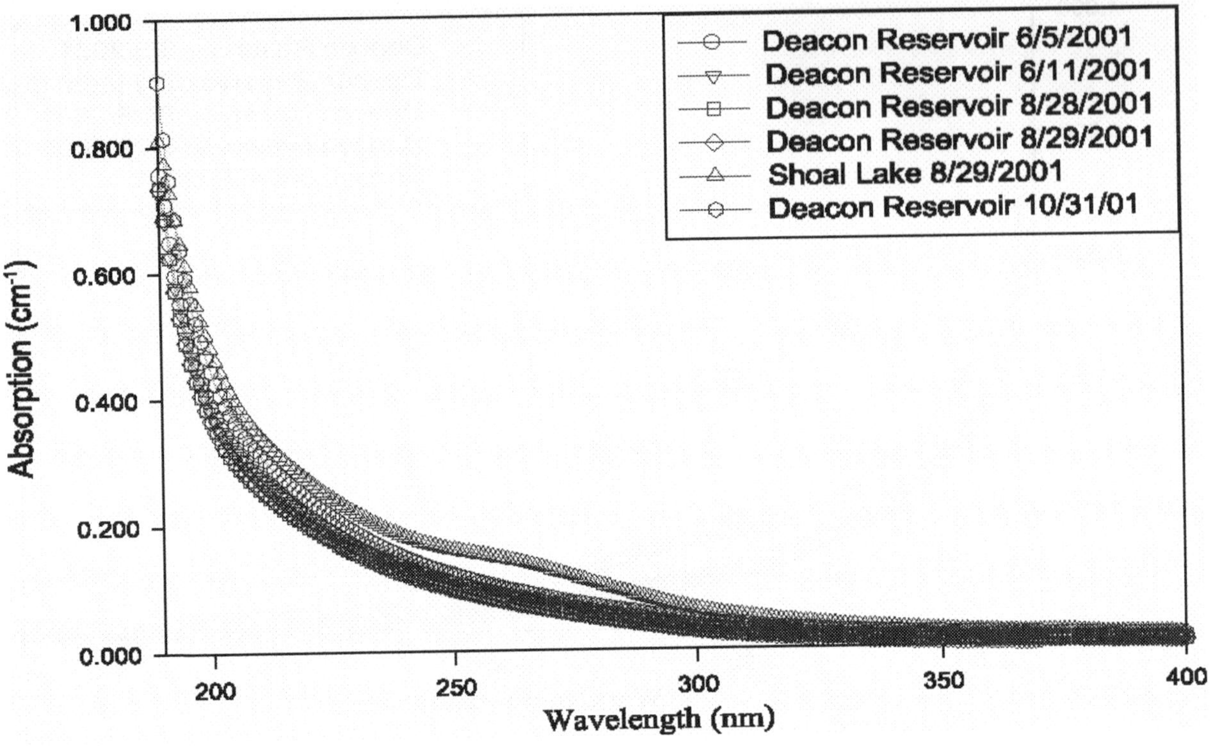

Figure 4.5 UV scans for Inactivation Study Nos. 1 through 3

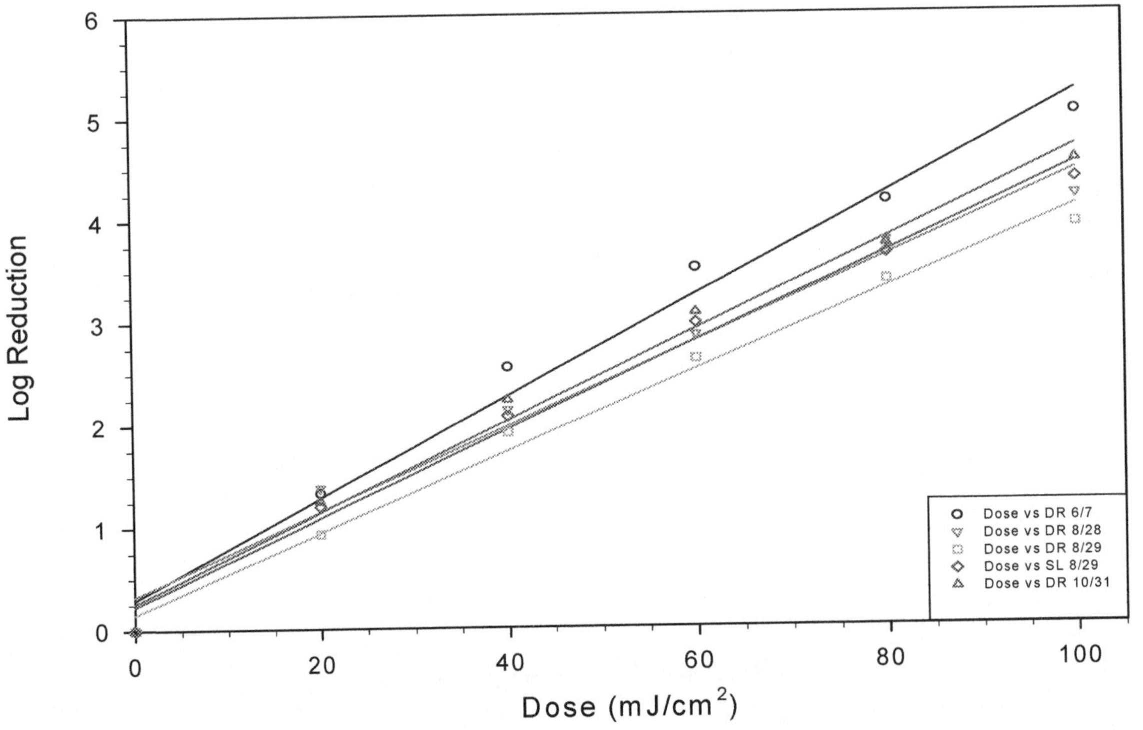

Figure 4.6 Collimated beam studies on Deacon Reservoir and Shoal Lake spiked with MS-2

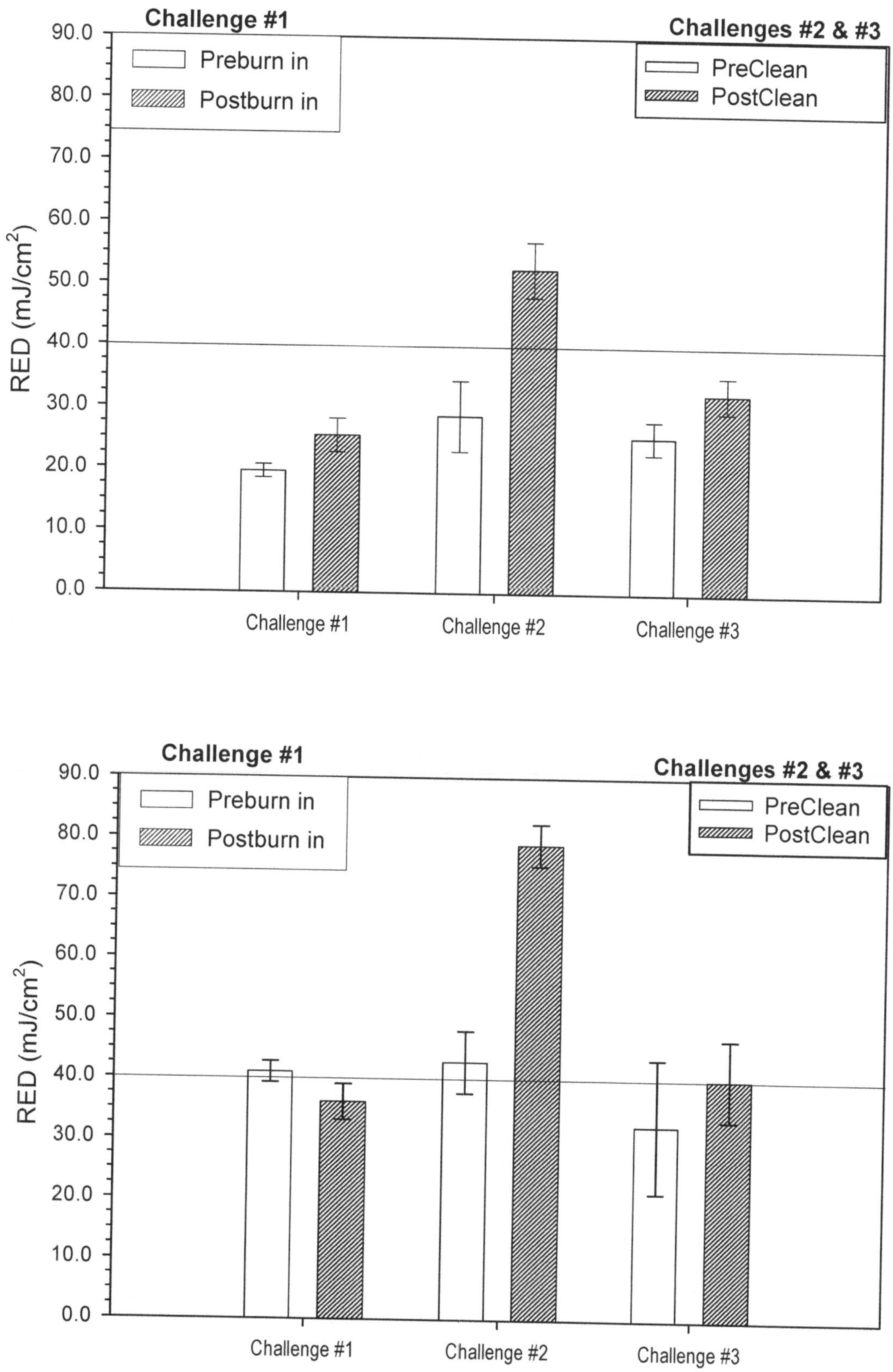

Figure 4.7 a, b pilot log reductions for Aquionics (above) and Wedeco (below) units

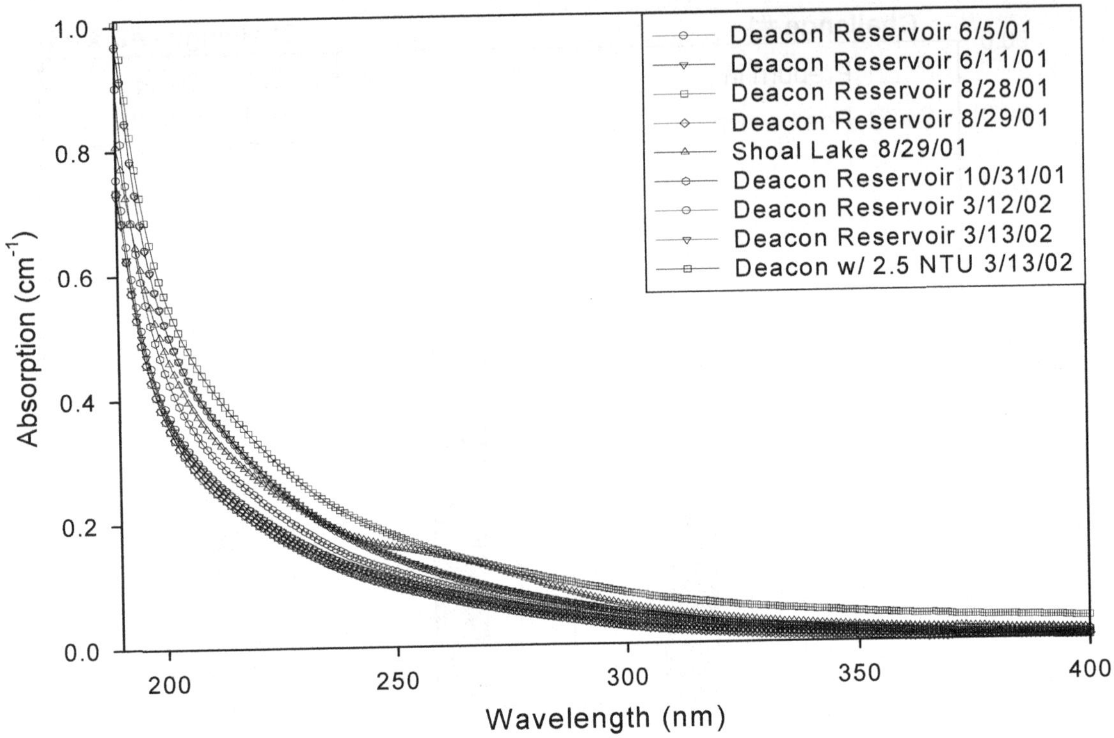

Figure 4.8 UV scans for Inactivation Test Nos. 1 through 4

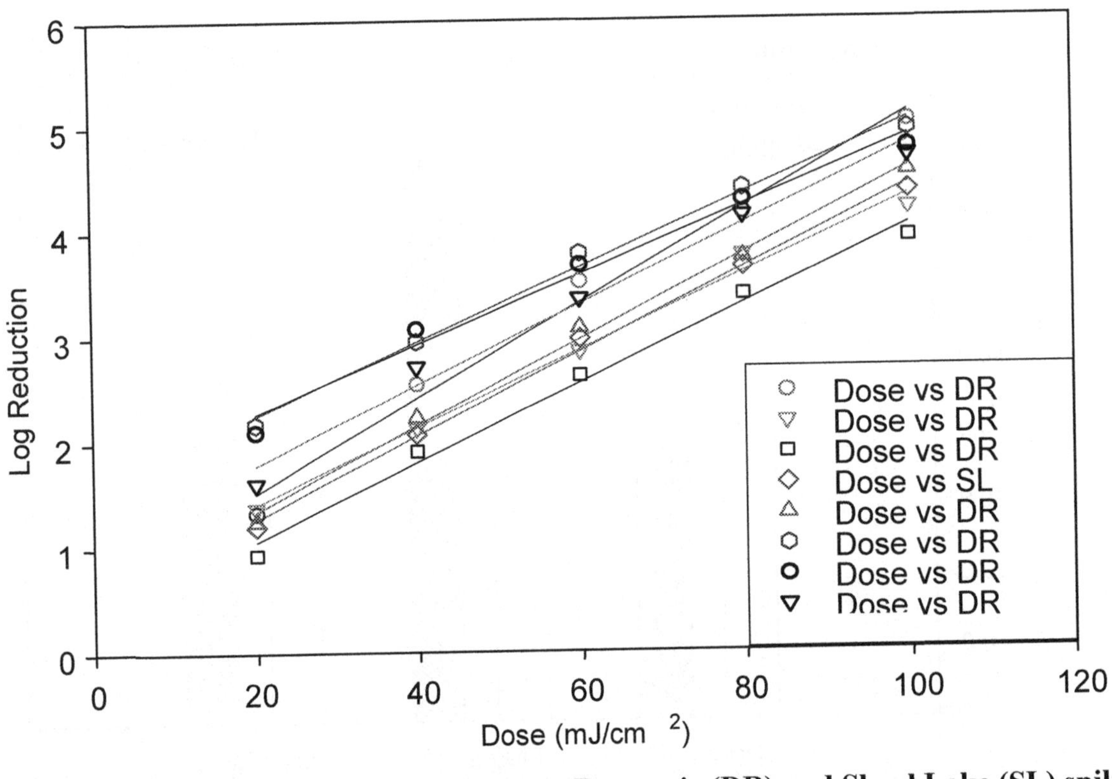

Figure 4.9 Collimated beam studies on Deacon Reservoir (DR) and Shoal Lake (SL) spiked with MS-2

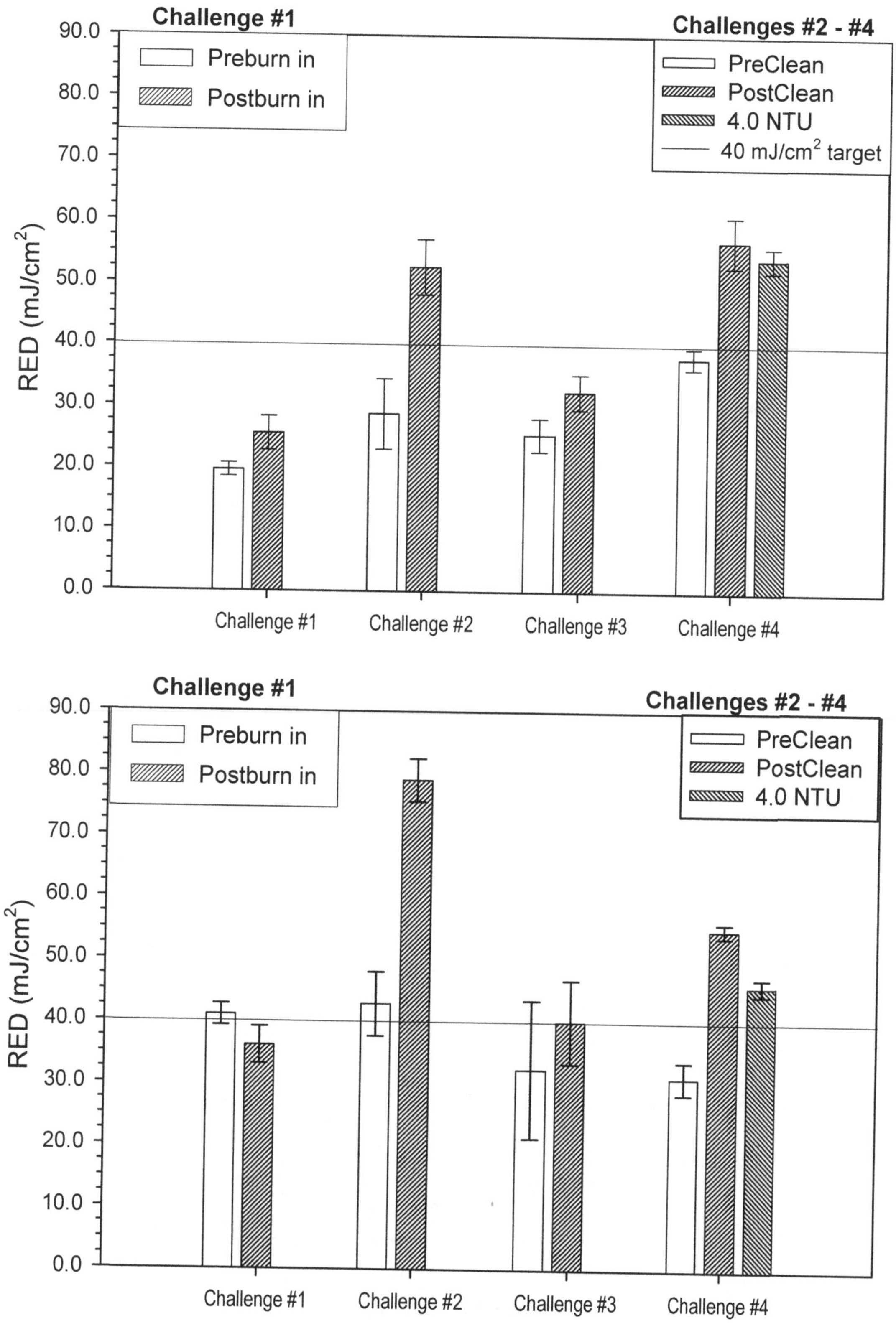

Figure 4.10 Pilot log reductions for Aquionics unit (above) and Wedeco unit (below)

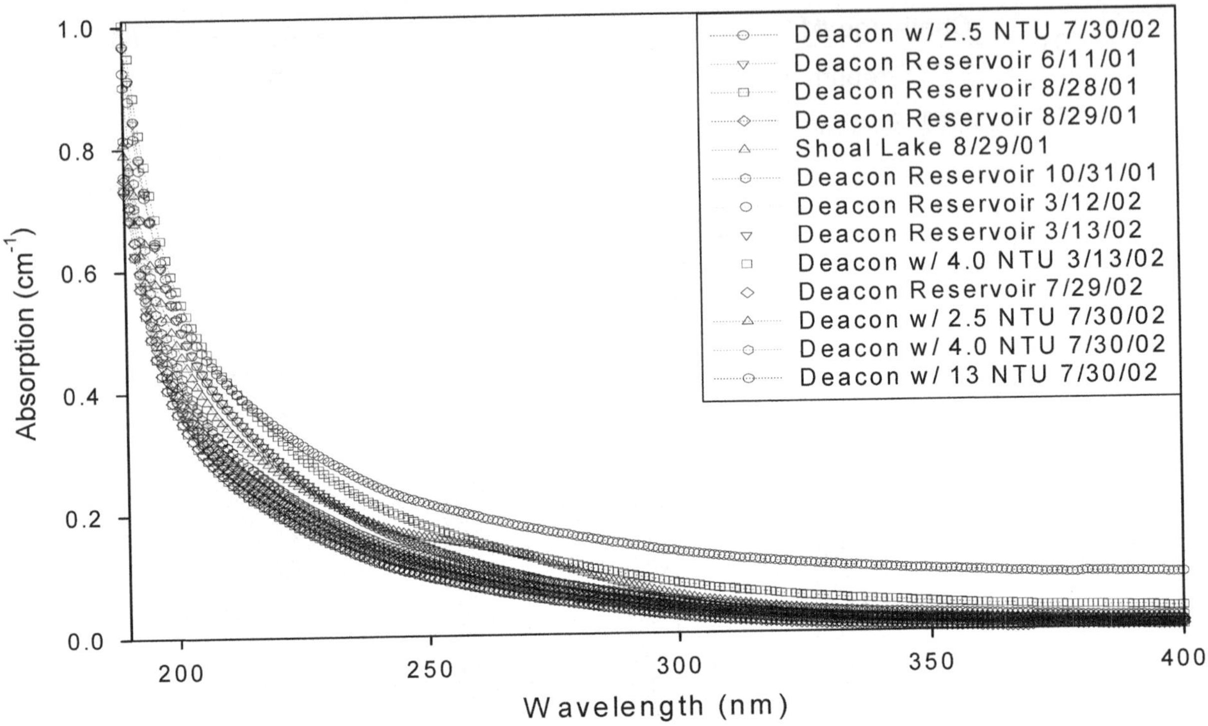

Figure 4.11 UV scans on waters from Winnipeg

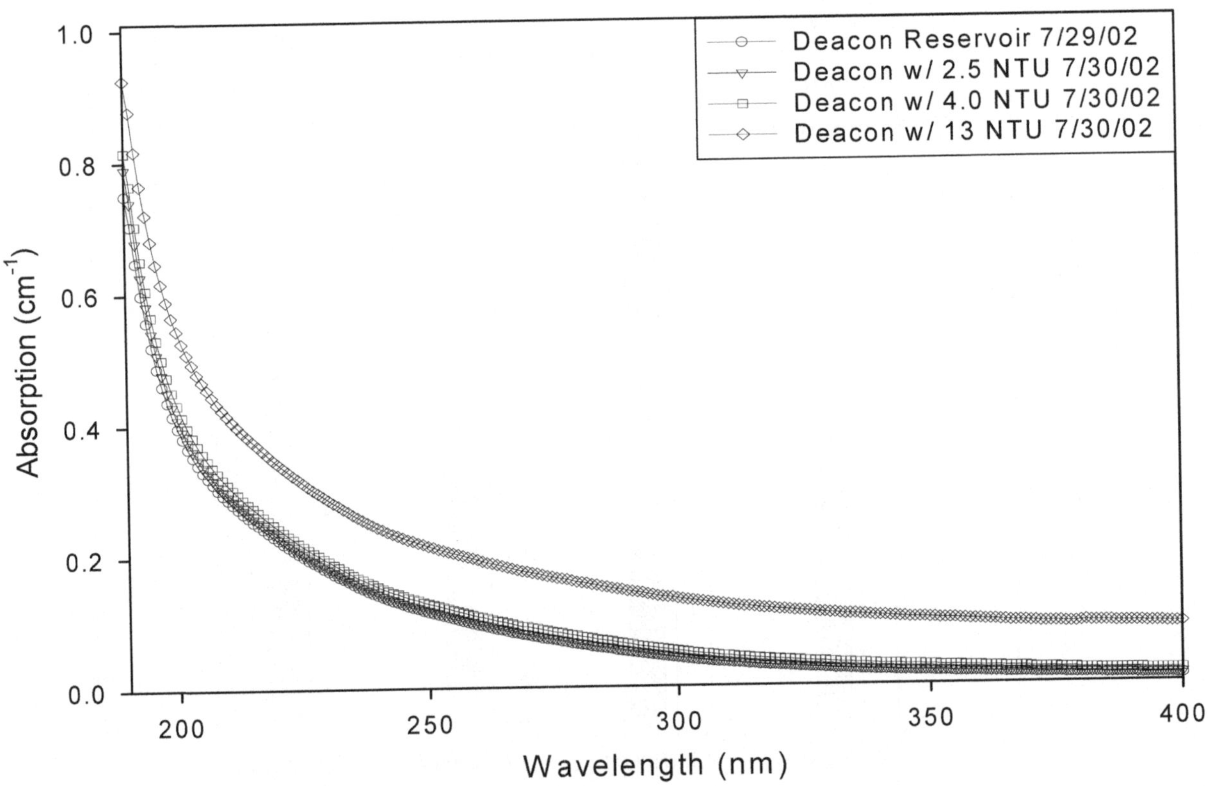

Figure 4.12 UV scans for Inactivation Study No. 5 at various turbidity levels

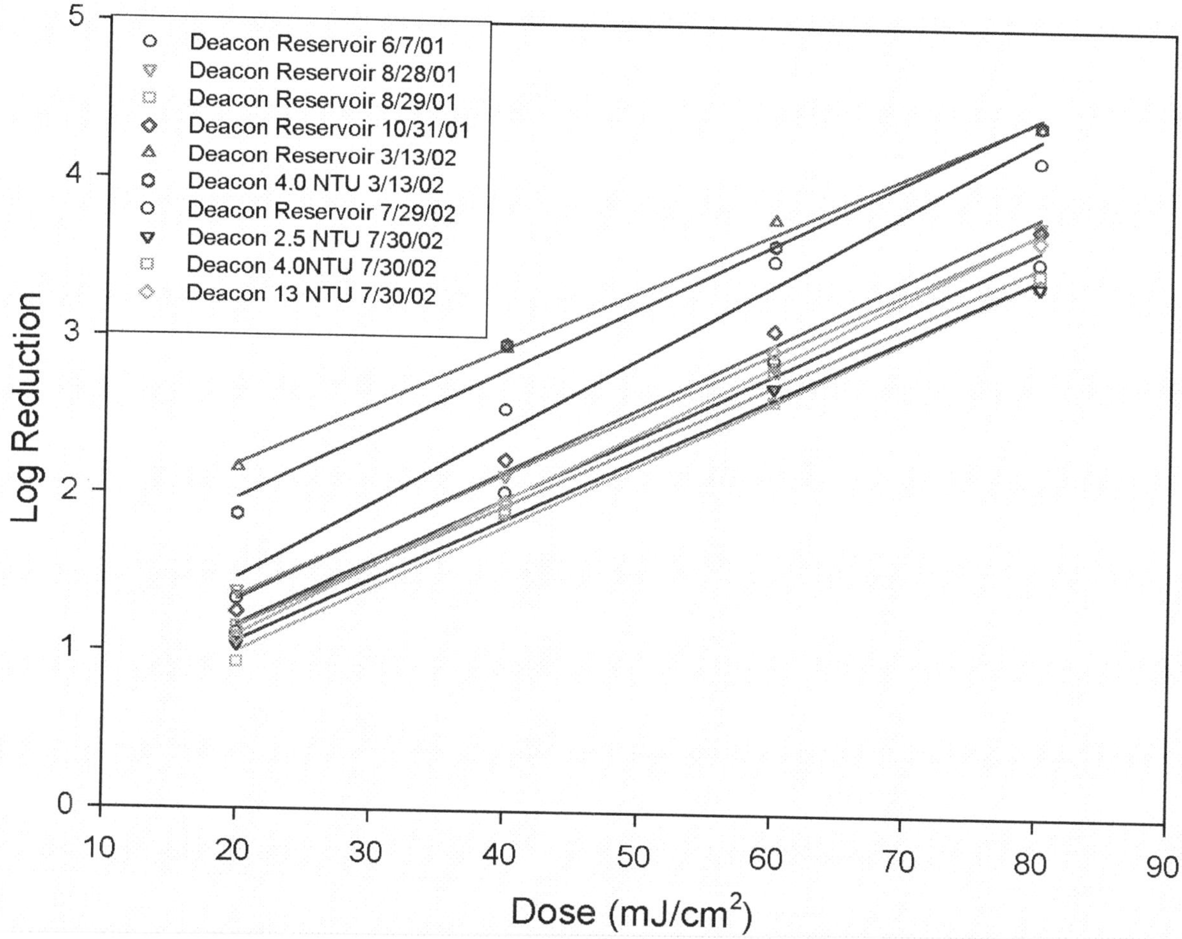

Figure 4.13 Collimated beam studies on Deacon Reservoir (DR) spiked with MS-2

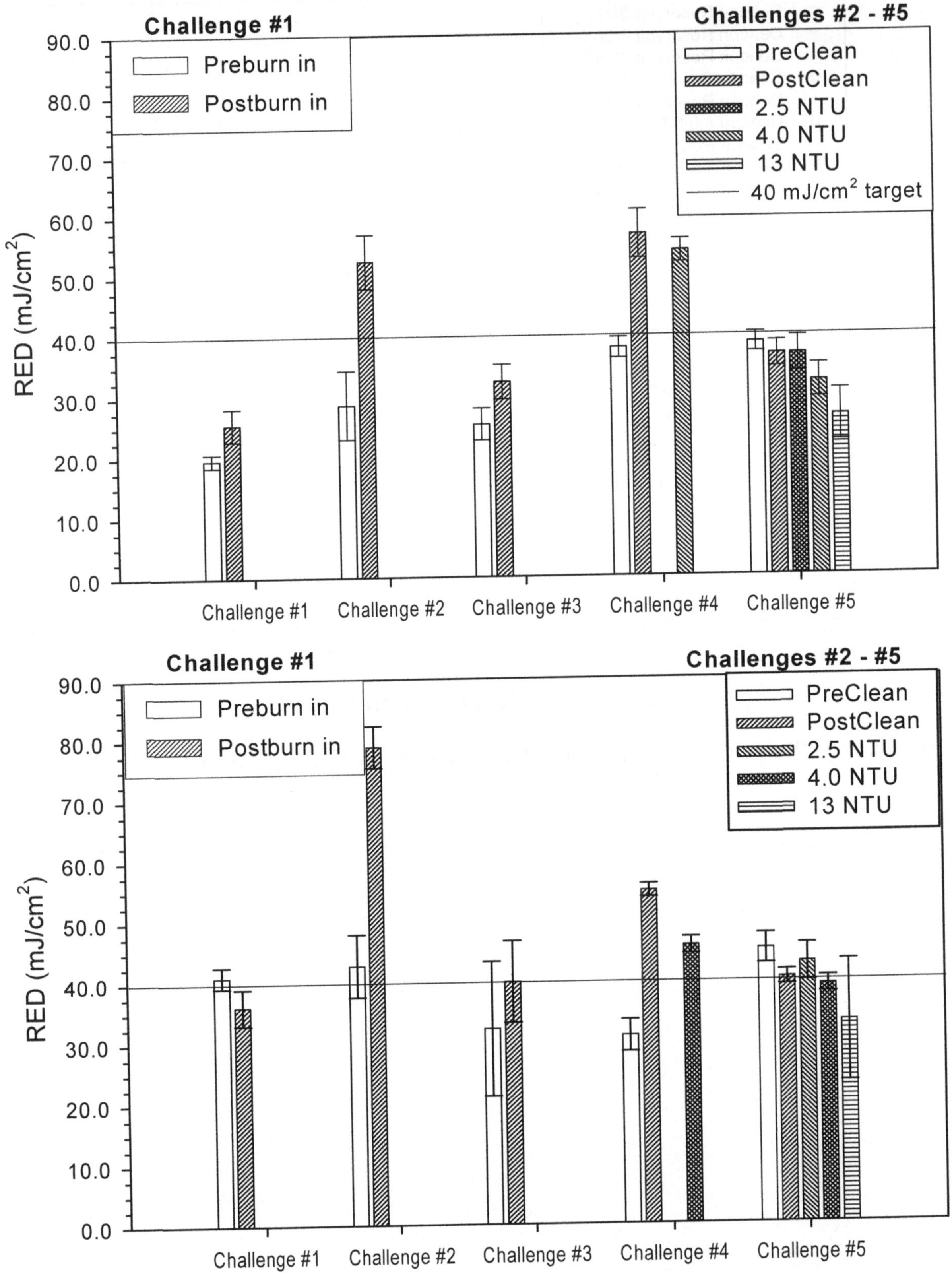

Figure 4.14 Pilot log reductions for Aquionics unit (above) and Wedeco unit (below)

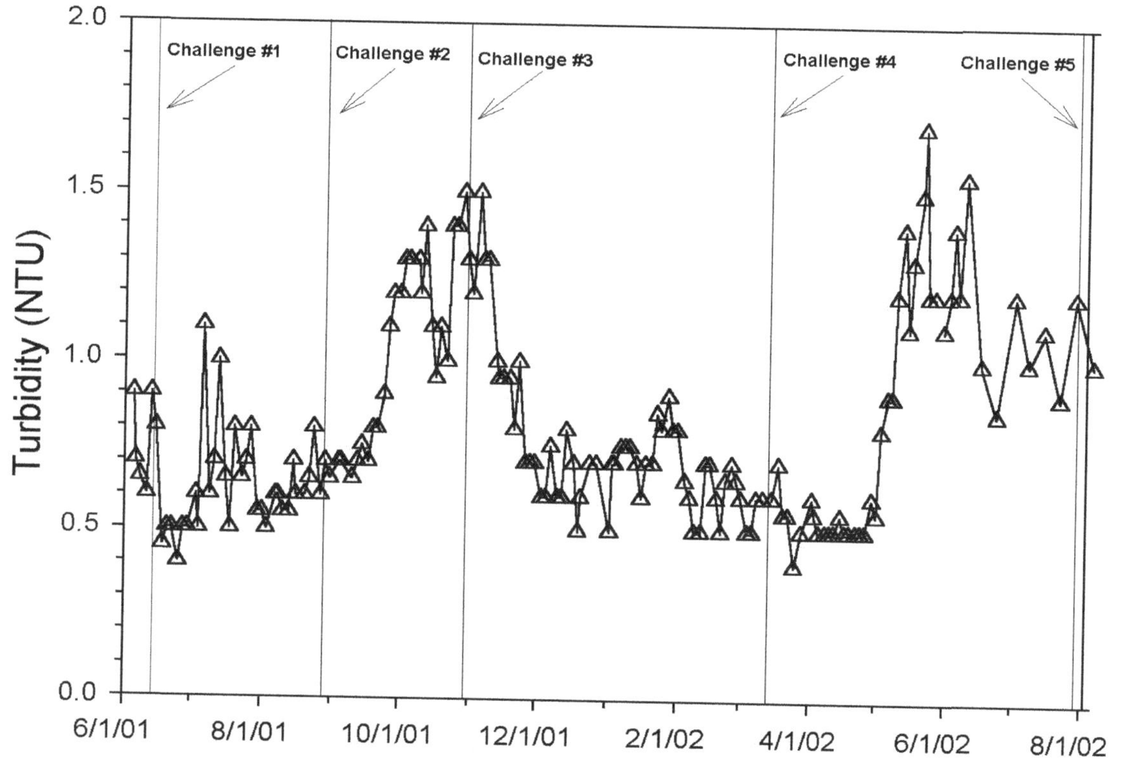

Figure 4.15 Turbidity profile for Deacon Reservoir water

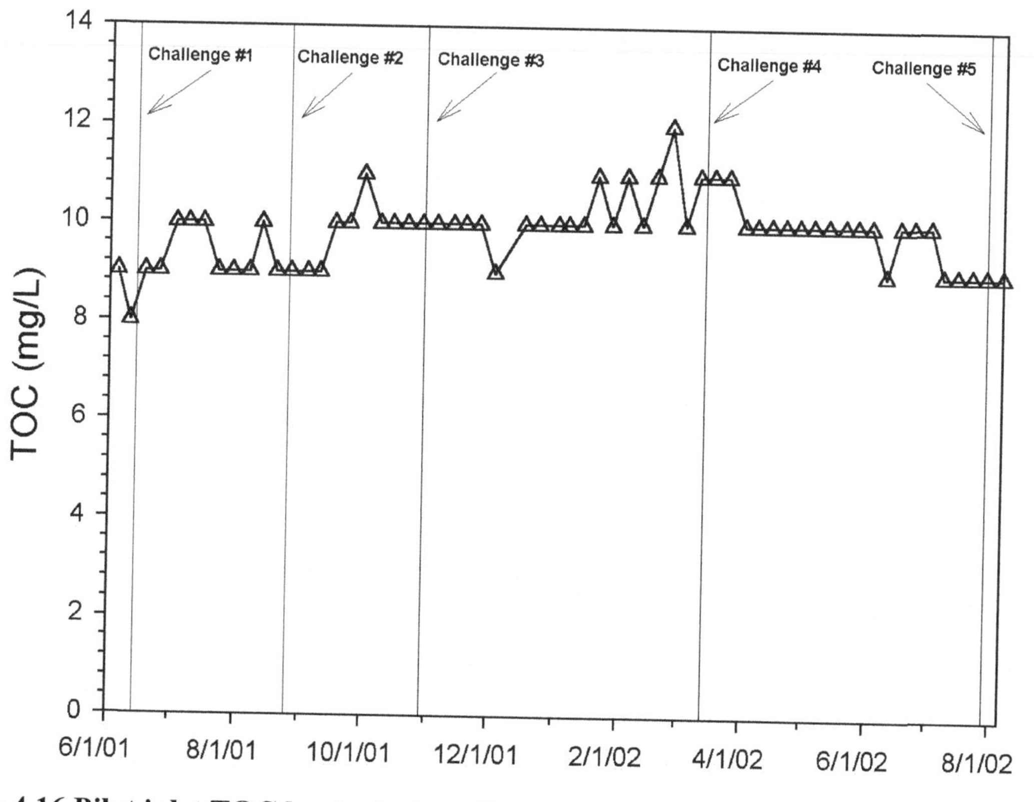

Figure 4.16 Pilot inlet TOC levels during pilot test program

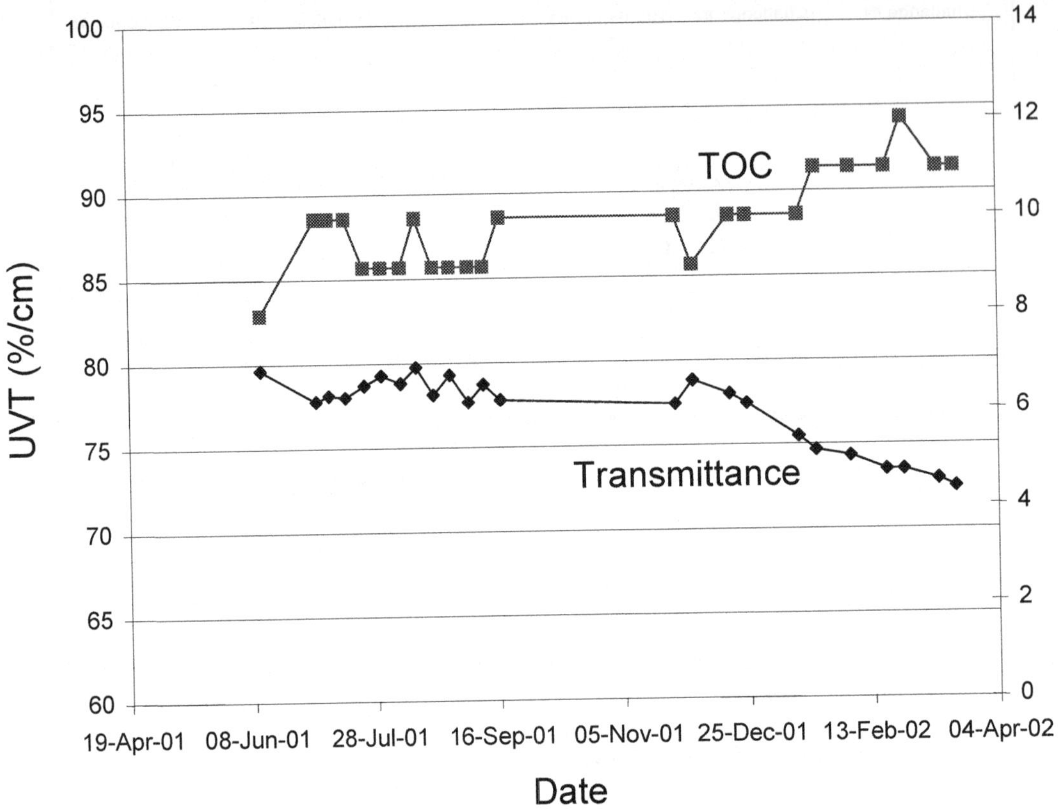

Figure 4.17 Raw water UVT versus TOC data

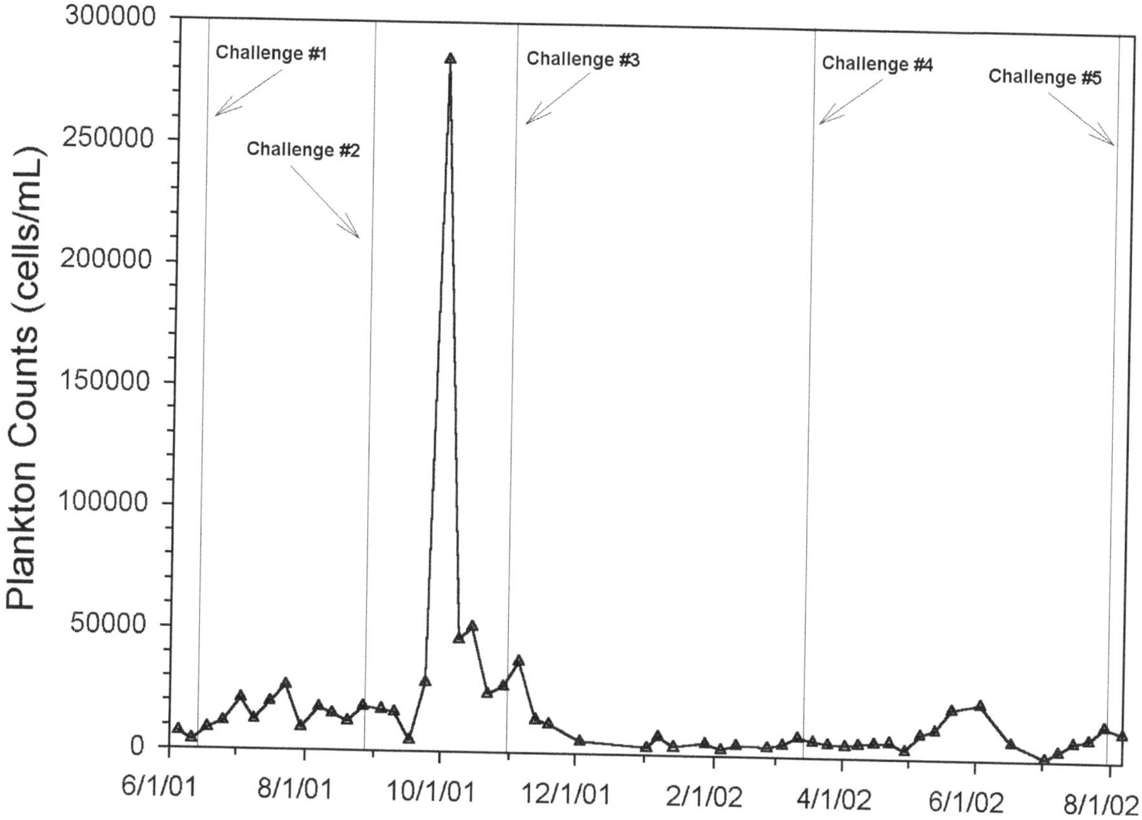

Figure 4.18 Raw water plankton count data

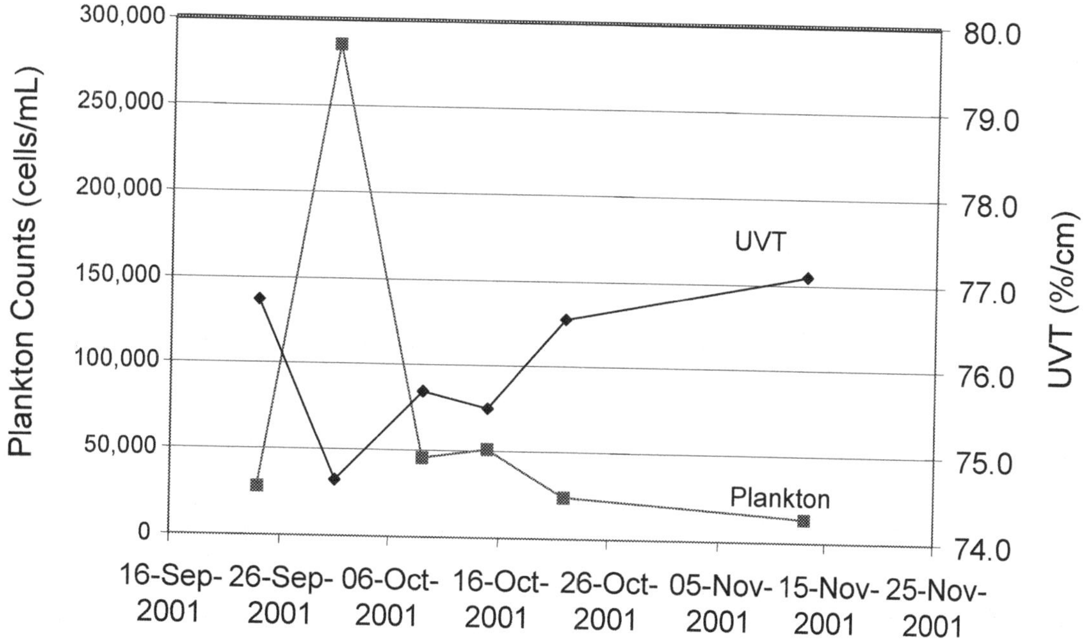

Figure 4.19 Plankton count versus UVT data

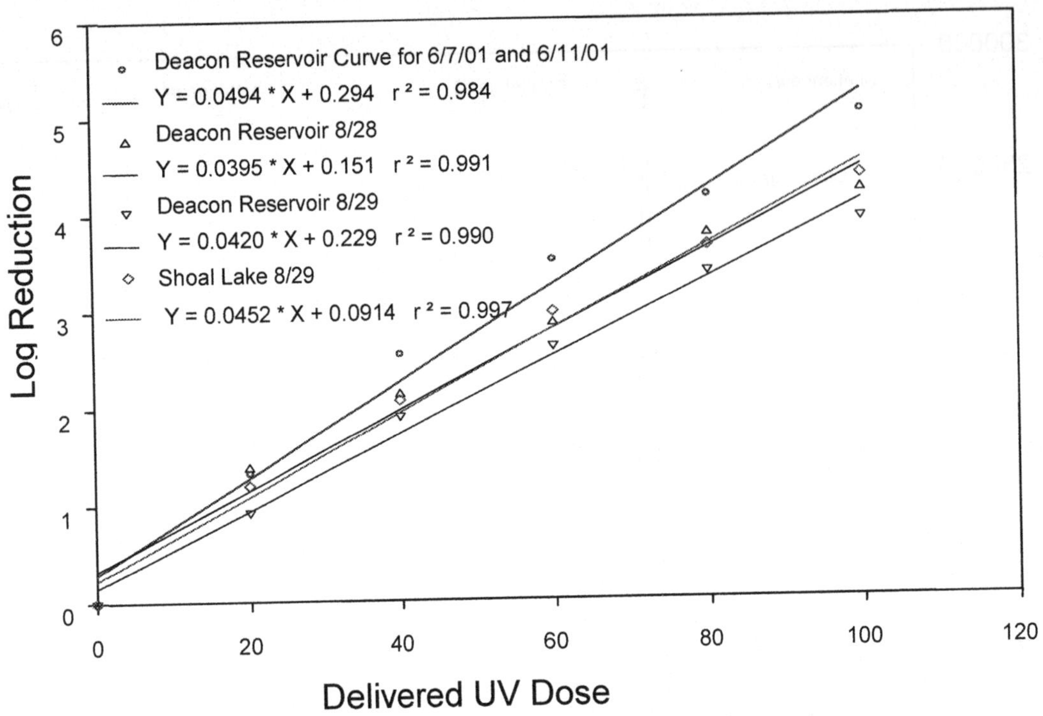

Figure 4.20 Collimated beam results for Inactivation Study Nos. 1 and 2 and Shoal Lake water

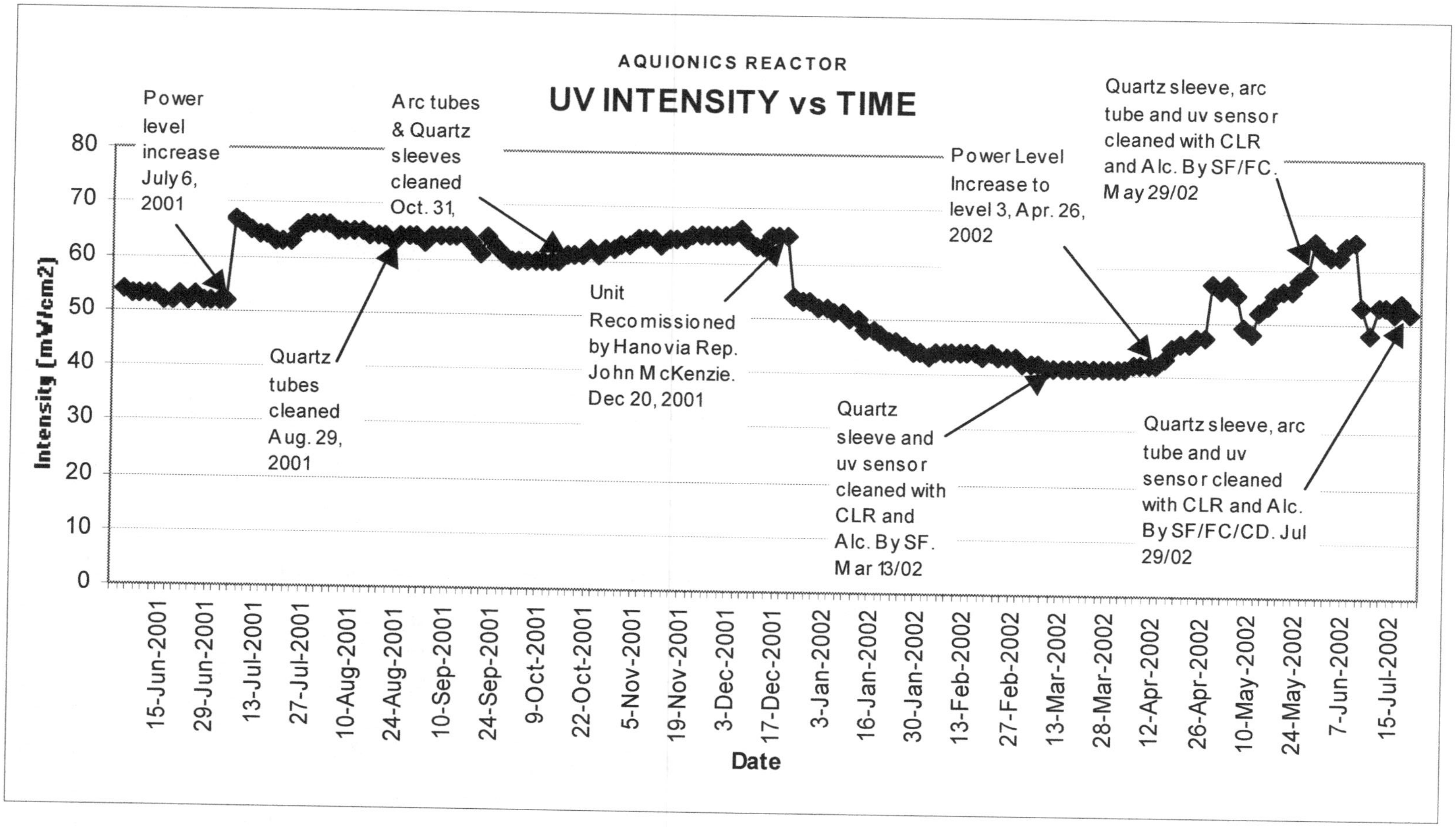

Figure 4.21 UV intensity data for Aquionics unit

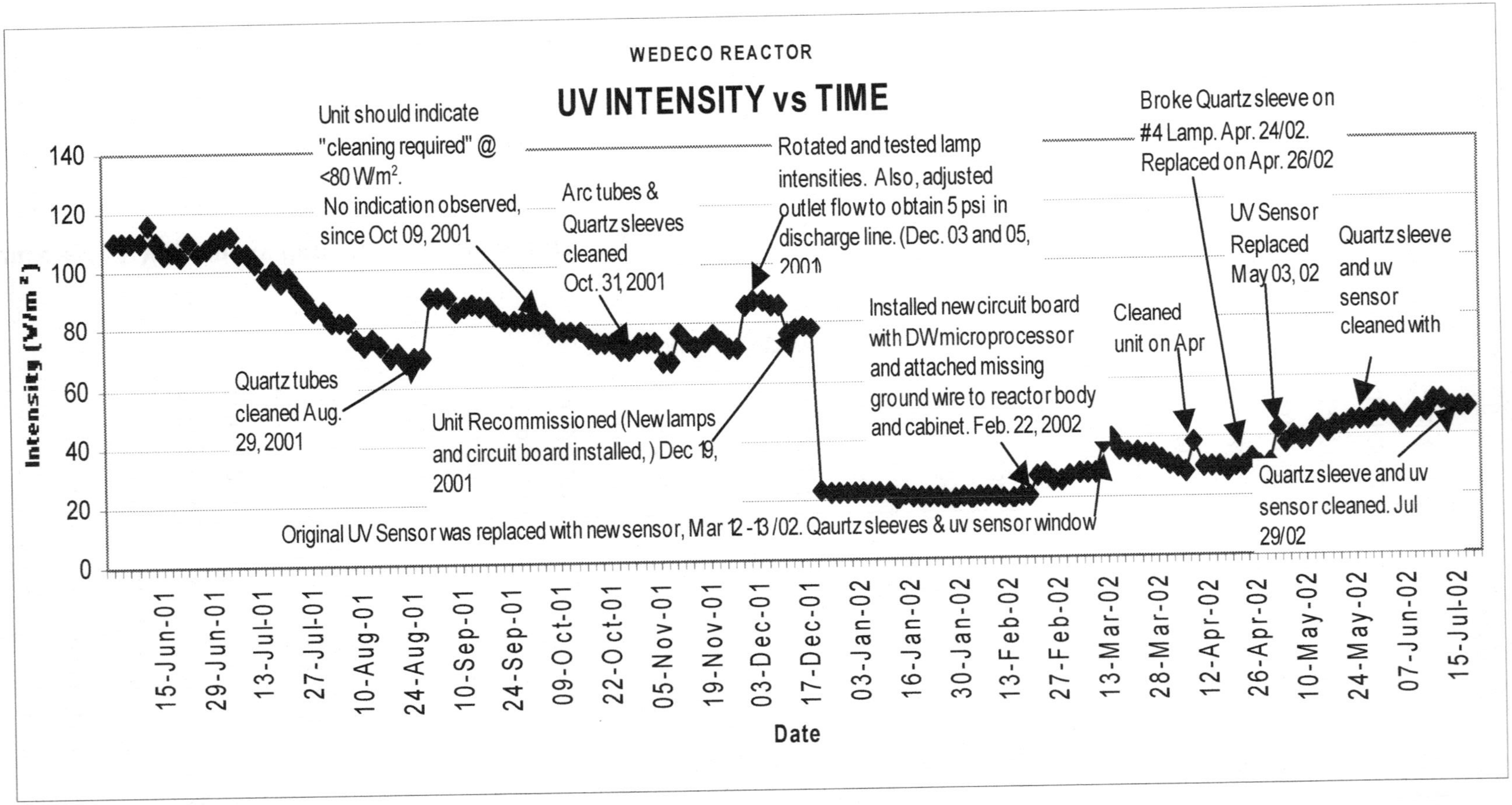

Figure 4.22 UV intensity data for Wedeco unit

CHAPTER 5
DISINFECTION AND DISINFECTION BYPRODUCT RESULTS AND ANALYSIS

SUMMARY

The specific objective of the disinfection and DBP studies was to:

- Determine the possible reduction and change in DBPs attainable after converting to UV for primary disinfection

The eight disinfection and DBP tests were designed to investigate changes in DBPs that can result from implementing UV primary disinfection and changing preoxidation and secondary disinfection practices.

DBP Test Nos. 1 Through 5

TTHM levels, as produced by the various disinfection scenarios, resulted in concentrations between 25 and 55 μg/L for prechloramination scenarios and 45 to 120 μg/L for those involving prechlorination. Actual THM values expected in the Winnipeg system may be different due to different levels of volatilization in the Deacon Reservoir. THM results and trends were consistent for DBP Test Nos. 1 through 5. THM levels did not change substantially after chloramination. Total haloacetic acids (THAA) were found in the final treated samples across a range of 25 to 55 μg/L when prechloramination was used. The corresponding values for prechlorination trains were 40–100 μg/L. Some biodegradation of dichloroacetic acid (DCAA) might have occurred in pretreatment, but this compound also showed long-term formation over the 5-day contact with chloramines

Aldehyde data showed a substantial increase due to the application of chlorine-based disinfectants. Final aldehyde levels were in the range of 25 to 55 μg/L. However, ketoacid levels were only moderately elevated. Final ketoacid levels were in the range of 10 to 40 μg/L. There seemed to be very little difference in the amount of nonchlorinated byproducts formed when comparing chlorination versus chloramination.

The DBP data showed some consistent qualitative trends that bear on the issue of process selection. First, there was no obvious difference between the UV/Cl_2 and Cl_2/UV sequences. Although some tests showed the sequence of Cl_2/UV to offer a modest advantage in controlling THM and HAA concentrations, other tests showed the reverse. Furthermore, the benefits in using one sequence versus the other were usually dampened or completely lost when viewing the final chloraminated data. Bioassay data (AOC and BDOC) also indicate that the sequencing of chlorine and UV is not important. Secondly, there was no clear indication that preformed chloramines led to reduced levels of DBPs as compared to in situ formed chloramines.

DBP Test Nos. 6, 7, and 8

On the basis of consistent and similar results of DBP Test Nos. 1 through 5, the DBP test program was modified to investigate other areas of interest with respect to disinfection and DBPs that may have potential impacts on Winnipeg's system. The modified DBP Test Nos. 6, 7, and 8 and objectives are summarized as follows:

- DBP Test No. 6 – UV impact on free-chlorine demand and disinfection byproduct formation.
  - To determine if UV causes an increase in free-chlorine demand and how much additional chlorine would have to be added.
  - If there is an increase in demand, determine if there are any detectable effects on DBP formation of this phenomenon.
- DBP Test No. 7 – Kinetics of disinfection byproduct formation from free chlorination of Shoal Lake Water.
  - To investigate various disinfection scenarios.
  - If free chlorination continued at headworks, test under different conditions, dose, temp, pH, etc.
- DBP Test No. 8 – UV impact on free-chlorine demand and emerging disinfection byproducts.
  - To determine potential NDMA levels in Shoal Lake and finished water. The USEPA has identified NDMA as a probable carcinogen. NDMA is currently regulated in California and Ontario.
  - To determine if the sequence of chlorination and UV might have any effect.
  - This DBP seems to be formed in chloraminating systems, and it is susceptible to decomposition by UV light.

For DBP Test No. 6, the data showed that at UV doses >400 mJ/cm^2, UV treatment caused a small increase in chlorine demand. The increase in demand was greater as UV dose increased. In general, DBPs also increased with increasing UV dose from 400 to 2000 mJ/cm^2. In some cases, these increases were as high as 20 percent over the control. At these higher UV doses, this test showed a trend indicating lower DBPs formation in the UV/Cl_2 sequence as compared to the reverse (Cl_2/UV).

For DBP Test No. 7, the partial DBP modeling for this run yielded the following equations for THMs and HAA5:

$$\text{THM } (\mu g/L) = 10^{-0.021} + (\text{Cl2 dose [mg/L]})^{0.444} + (\text{Time [hrs.]})^{0.191} + (\text{pH})^{1.518}$$
$$\text{HAA5 } (\mu g/L) = 10^{1.771} + (\text{Cl2 dose [mg/L]})^{0.718} + (\text{Time [hrs.]})^{0.128} + (\text{pH})^{-0.578}$$

DBP formation increases with chlorine dose and contact time under all pH conditions. THM concentration increased with pH, while HAA5 concentration decreased with pH at all doses and contact times. The general trends for the kinetic data for this water were generally in agreement with those reported for other waters in the literature.

For DBP Test No. 8, there was no obvious difference between the UV/Cl_2 and Cl_2/UV sequences with respect to the THM and HAA formation. However, this run showed a clear trend at UV doses > 400 mJ/cm^2 indicating lower TOX and NDMA formation in the Cl_2/UV sequence as compared to the reverse (UV/Cl_2). UV treatment caused a noticeable increase in chlorine demand. The increase in demand was greater as UV doses were increased from 400 to

2000 mJ/cm^2. DBPs concentrations also increased substantially with increasing UV dose. This was especially noticeable with TOX formation.

INTRODUCTION

The disinfection and DBP bench-scale testing was designed to investigate changes in DBPs that can result from implementing UV primary disinfection and changing preoxidation and secondary disinfection practices. The disinfection bench-scale studies were conducted at UMass. The bench-scale testing provided information on the impacts of proposed changes in disinfection practices at Shoal Lake and Deacon Reservoir on the DBP levels expected in Winnipeg's system. These results are applicable to other potable water systems, especially those with relatively high organics and algae concentrations. The bench-scale DBP testing was divided into four main areas of investigation:

- Water quality characterization
- Disinfectant demand tests
- Disinfection byproduct testing
- Testing for biodegradable organic matter (BOM)

Eight DBP tests were included in the study. Test protocols, materials, and methods are outlined in Chapter 3. This chapter summarizes the results of DBP Tests 1 through 8. Due to the similarity of results obtained during the first five DBP tests, the DBP test plan was modified for the last three DBP tests. Consequently, DBP Test Nos. 1 through 5 will be discussed as a group, while DBP Test Nos. 6, 7, and 8 will be discussed individually.

DBP TEST NOs. 1 THROUGH 5

Water Quality Characterization

Table 5.1 summarizes specific water quality data for DBP sample sets 1 through 5. A key water quality parameter is the specific UV absorbance (SUVA), which is equal to the UV absorbance at 254 nm (cm^{-1}), divided by the DOC (mg/L) and multiplied by 100. SUVA is an indication of the organic content of the water and the potential for the formation of DBPs.

Table 5.1
Raw water quality for DBP Test Nos. 1 through 5

DBP test no.	Date	Temp °C	pH	UV 254 (cm-1)	TOC (mg/L)	SUVA (L/mg/m)
1	6-June-01	15.5	7.6	0.164	12.80	1.28
2	16-July-01	22.5	7.9	0.111	7.94	1.51
3	10-Oct-01	11.0	7.0	0.150	9.49	1.58
4	26-Nov-01	2.0	7.5	0.148	9.45	1.57
5	30-Jan-02	2.0	7.0	0.159	8.69	1.63

Aside from DBP Test No. 1, all SUVA values are consistent, ranging from 1.51 to 1.63 L/mg/m. These low SUVA values are generally associated with low-specific DBP precursor levels, but this water is unusual in that it has both a low SUVA and a high DOC. This indicates that the raw water has a slightly higher humic character but still quite low for waters with this level of TOC. The research team does not have much experience with such waters that would allow the team to make judgments on the reasons at this point.

Disinfectant Demand Tests

Chlorine Dioxide

In keeping with established protocol, the raw water was tested with a 1.5 mg/L ClO_2 dose. This level was selected because about 70 percent of the chlorine in consumed chlorine dioxide ends up as chlorite. Therefore, if chlorite levels must be kept below the MCL of 1 mg/L, chlorine dioxide losses of more than 1.5 mg/L cannot be sustained. Results showed that there was no measurable chlorine dioxide residual at the end of 24 hours. This indicated that ClO_2 could not be used with this water to maintain a lasting residual without exceeding a 1 mg/L chlorite standard. Consequently, no additional work was done with chlorine dioxide as a preoxidant.

Chlorine/Chloramine

One of the first experiments was an oxidant demand study using free chlorine and chloramines (both in situ formed and preformed). Dose versus residual curves were developed so that the appropriate dose to achieve the desired residual (0.2 mg Cl_2/L) at the end of the preoxidation contact time (30 hours) could be estimated. Typical curves for DBP Test No. 1 are presented in Figures 5.1 and 5.2. Similar curves were developed for DBP Test Nos. 2 through 5.

The doses that were adopted for bench-scale pretreatment were based on these demand curves. The values chosen for DBP Test No. 1 are presented in Table 5.2. This procedure was carried out for each of the DBP Test Nos. 1 through 5.

Chlorine Residuals

At each important experimental juncture, chlorine residuals were measured using the DPD method. This includes the measurement of free chlorine, monochloramine (monoCLM) and dichloramine (diCLM).

The sample key numbers (No. 1 to No. 29) represented in Figure 5.3 correspond to the different points for which chlorine data are available in the 8- and 20-day pretreatment tests. Twenty-two of these represent actual chlorine residual measurements. However, seven of these (Nos. 2, 3, 4, 8, 9, 10, and 20) are reserved for dosed values, which should reflect the concentrations at the instant of chlorine addition.

Table 5.2
Results of preliminary oxidant tests for DBP Test No. 1

Oxidant	Dose (mg Cl_2/L)	Target residual (mg/L)
Free chlorine	5.25	0.2
In situ formed chloramines	0.55	0.2
Preformed chloramines	0.50	0.2

These seven are from dose/residual tests and correspond to:

- Pretreatment doses (Nos. 2, 3, and 4).
- Intermediate treatment: doses to reach the desired contact time (CT) value after 20 minutes (Nos. 8, 9, and 10). The particular CT value for these tests was 1.9 min.*mg/L This leads to a desired free residual after 20 minutes of about 0.1 mg/L.
- Post-treatment dose: to achieve the desired residual at the end of the 5-day post-treatment period (No. 20).

Figures 5.4 to 5.8 present 8- and 20-day chlorine dose and residual data for DBP Tests 1 to 5. For all the 8-day tests, the chlorine residual was completely depleted at the end of the 8-day pretreatment period. This is to be expected as the preoxidant dosing criteria called for a small (0.2 mg/L) residual after 30 hours.

The first-stage intermediate samples (Nos. 11 through 13) and the final intermediate samples (Nos. 15, 17, and 19) had FRCs close to 0.14 mg/L as specified in the protocol. The intermediate chlorine residuals had higher than expected combined residuals. Based on the protocol, these should have no detectable combined residual. The presence of a substantial chloramine residual in these samples may be due to the oxidation of organic amines present in the water.

Final chloramine doses were estimated based on a short-term chlorination test, because a full 5-day test would have resulted in undesirable sample aging. For DBP Test Nos. 1 and 2, the final chloramine dose (1.5 mg/L) was a bit higher than desirable because all of the final samples had combined residuals above the 0.6 mg/L target value. For DBP Test No. 3, the final chloramine dose (1.1 mg/L) was appropriate because all of the final samples had combined residuals close to 0.6 mg/L target value. For DBP Test No. 4, the final chloramine dose (1.1 mg/L) was near optimal because all of the final samples had combined residuals close to 0.6 mg/L target value. For DBP Test No. 5, the final chloramine dose (1.1 mg/L) was marginally high because most of the final samples had combined residuals slightly above the 0.6 mg/L target value.

From the 20-day data, it can be seen that the chlorine residual was completely depleted at the end of the 20-day pretreatment period (sample Nos. 5 through 7). Again, this is to be expected as the preoxidant dosing criteria called for a small (0.2 mg/L) residual after only 30 hours. Typical first-stage intermediate samples (Nos. 11 through 13) were found to contain nearly the desired level of free residual but a surprisingly large combined residual. The final intermediate samples (Nos. 14 through 19) also had higher than expected combined residuals. Small free residuals were seen in the samples treated with UV before chlorine (Nos. 15, 17, and 19). However, there was no measurable free residual in the samples treated with UV after chlorine (Nos. 14, 16, and 18), reflecting the ability of UV treatment to destroy the FRC. Again, the presence of a substantial chloramine residual in these samples suggests that either the prior chlorine dosing was too low or that there was a slow release of ammonia from the natural organic matter. Because the typical ammonia levels in the raw water were between 0.08 and 0.13 mg/L, a slow release of ammonia from natural organic matter may have been occurring.

Disinfection Byproduct Testing

The sampling key for DBPs is exactly the same as for the chlorine doses and residuals (Figure 5.3).

THM Tests

Figures 5.9 through 5.13 contain all of the THM data for these tests.

THMs – DBP No. Test 1, Figure 5.9. For the 8-day test, as expected the raw water (sample No. 1) had almost no THMs. This was consistent for all the tests. The intermediate treatment THM levels (sample Nos. 14 and 15) were lower than expected for the prechlorination train (sample No. 5). It was suspected that this was due to THM volatilization in the lab-scale UV reactor. Experiments verified this, and the protocols were adjusted to try to minimize this effect; however, it could not be eliminated. Samples 7 through 10 show almost identical THM levels, indicating that the sequence of UV and intermediate chlorination is not very important to THM formation. Post-treatment THM levels are similar to the intermediate levels. This is not unexpected because a second stage of chloramination normally does not have a significant impact on THM concentrations. All of the final treated samples had THM levels in the range of 30 to 40 µg/L.

For the 20-day test, of the pretreated samples, only the prechlorinated sample (sample No. 5) had a substantial THM level. The 20-day pretreated THM levels were significantly higher than the 8-day levels (190 versus 140). Such changes are surprising. However, some increase might occur due to slow hydrolysis that continues even in the absence of residual chlorine. This was unexpected due to the chlorine residual being depleted after 8 days. The other two samples (sample Nos. 6 and 7) were below 10 µg/L, with the preformed-chloramine sample slightly lower than the in situ chloramine sample. Once again, the intermediate treatment THM levels (sample Nos. 14 and 15) were lower than expected for the prechlorination train (sample No. 5). The likely reason is volatilization, as discussed previously. Sample Nos. 7 through 10 show almost identical THM levels, again indicating that the sequence of UV and intermediate chlorination is not very important to THM formation. Post-treatment THM levels indicate no increase over the intermediate levels. All of the final treated samples had THM levels in the range of 40 to 45. This is higher than the levels seen with the 8-day pretreatment (i.e., 30 to 40 µg/L). It is unclear why this would be the case because the only difference is in the length of initial exposure to chloramines. Perhaps some reactions are occurring that make the NOM more susceptible to DBP formation when free chlorine is applied.

THMs – DBP No. Test 2, Figure 5.10. For the 8-day test, of the pretreated samples, only the prechlorinated one (sample No. 5) had a high THM level. The other two were between 15 and 30 µg/L, with the preformed-chloramine sample slightly lower than the in situ chloramine sample. These results are qualitatively similar to those in DBP Test No. 1. However, the magnitude of THM formation in the prechloraminated samples is several times larger than that in

DBP Test No. 2. The intermediate treatment THM levels were much lower than expected for the prechlorination train (sample Nos. 14 and 15). Again, it was suspected that this resulted from volatilization in the lab-scale UV reactor. UV destruction was also considered, although this seems less likely.

Experiments conducted around the time of DBP Test No. 2 seemed to verify volatilization in the lab-scale UV reactor. Sample Nos. 16 and 18 exhibited similar levels, and these were higher than those for sample Nos. 17 and 19. This is true even though sample Nos. 16 and 18 had slightly lower chlorine residuals. This would suggest that the intermediate treatment sequence of UV followed by chlorination produces lower levels of THMs rather than the reverse. Results from DBP Test No. 1 failed to show any obvious difference in this sequence. As expected, post-treatment THM levels reflect the intermediate levels. The only possible contrary data are those for the prechlorine train. Here, the final concentrations seemed to increase by as much as 20 μg/L with the Cl_2/UV sequence, but there was very little increase for the UV/Cl_2 sequence. All the final treated samples that were not prechlorinated had THM levels in the range of 25 to 38 μg/L. The advantage that was seen for the UV/Cl_2 sequence in the preformed-chloramine train (No. 17 versus No. 16) is not apparent in the final samples (No. 26 versus Nos. 24 and 25). However, the in situ chloramine train still showed a benefit for this sequence (No. 29 versus Nos. 27 and 28).

For the 20-day test, of the pretreated samples, only the prechlorinated sample (sample No. 5) had a substantial THM level, and this level was nearly the same as that observed for the 8-day test (about 105 μg/L). This was expected as chlorine residual was depleted after 8 days and further THM formation should be discontinued at this point. The other two were below 25 μg/L, with the preformed-chloramine sample still slightly lower than the in situ chloramine sample. Once again, the intermediate treatment THM levels were lower than expected for the prechlorination train (sample Nos. 14 and 15). The likely reason is volatilization, as discussed previously.

As with the 8-day samples, here, the UV/Cl_2 sequence shows much lower THM levels than the reverse (Nos. 17 versus 16, and 19 versus 18). This would suggest that the intermediate treatment sequence of UV followed by chlorination is superior with respect to THM formation. Results from DBP Test No. 1 failed to show any obvious difference in this sequence. Post-treatment THM levels showed increases for the prechlorine train. As with the 8-day data, the final concentrations seemed to increase by as much as 20 μg/L with the Cl_2/UV sequence, but there was very little increase for the UV/Cl_2 sequence. All of the final treated samples that were not prechlorinated had THM levels in the range of 25 to 40 μg/L. The advantage that was seen for the UV/Cl_2 sequence in the preformed-chloramine train (No.17 versus No. 16, and No. 19 versus No. 18) was not apparent or even reversed in the final samples (No. 26 versus Nos. 24 and 25, No. 29 versus Nos. 27 and 28).

THMs – DBP No. Test 3, Figure 5.11. For the 8-day test, of the pretreated samples, only the prechlorinated sample had a high THM level. The other two samples were approximately 10 μg/L. These results are qualitatively similar to those in DBP Test Nos. 1 and 2. The intermediate treatment THM levels were a bit lower than expected for the prechlorination train (sample Nos. 14 and 15). As indicated previously, this results from volatilization in the lab-scale UV reactor. Sample Nos. 16 and 18 exhibited similar levels. These levels were slightly lower than those for sample Nos. 17 and 19. This indicates no obvious difference in the treatment sequence of chlorine and UV. Post-treatment THM levels reflect the intermediate levels. For the prechlorination train, the final THM levels were similar to those of the intermediate THM, which showed that postchloramination in the prechlorination train added almost no THMs. All of the final treated samples that were not prechlorinated had THM levels in the range of 30 to 45 μg/L. The small advantage that was seen for the Cl_2/UV sequence in the preformed-chloramines train (No. 16 versus No. 17) and in situ chloramine train (No. 18 versus No. 19) appeared again in the final samples (Nos. 24 and 25 versus No. 26; Nos. 27 and 28 versus No. 29). The preformed-chloramine train and in situ chloramine train showed similar THM concentrations during this series of treatments.

For the 20-day test, of the pretreated samples, only the prechlorinated sample had a substantial THM level (about 140, and this level was a bit higher than that observed for the 8-day test [about 125 μg/L]). The other two samples were below 10 μg/L. Once again, the intermediate treatment THM levels were lower than expected for the prechlorination train (sample Nos. 14 and 15). The likely reason is volatilization, as discussed previously. As with the 8-day samples, there was no obvious difference between the different UV and Cl_2 sequences. Post-treatment THM levels showed a little increase for the prechlorine train. The final concentrations seem to have increased by as much as 10 μg/L with the Cl_2/UV sequence, but there was no increase for the UV/Cl_2 sequence. All of the final treated samples that were not prechlorinated had THM levels in the range of 40 to 55 μg/L. As with the 8-day test, the Cl_2/UV/preformed chloramine (CLM) and Cl_2/UV/IS CLM sequences showed a modest advantage with a 5 to 20 percent reduction over the UV/Cl_2/IS CLM sequence (Nos. 24 and 25 versus No. 26; Nos. 27 and 28 versus No. 29).

THMs – DBP Test No. 4, Figure 5.12. For the 8-day test, of the pretreated samples, only the prechlorinated sample had a high THM level. This was about 65 μg/L, which is substantially lower than in previous tests, probably because of the low temperature (2°C). The other two pretreated samples were approximately 10 μg/L. These results are qualitatively similar to those in DBP Test Nos. 1, 2, and 3. The intermediate treatment THM levels were slightly lower than expected for the prechlorination train (sample Nos. 14 and 15), again due to volatilization in the lab-scale UV reactor. Sample Nos. 16 and 18 exhibited similar THM levels to sample Nos. 17 and 19. This indicates no obvious difference in the treatment sequence of chlorine and UV. Post-treatment THM levels reflected the intermediate levels. For two of the three prechlorination train samples, the final THM levels were slightly higher than those of the intermediate THM, which showed that postchloramination in the prechlorination train added little or no THMs. All but one of the final treated samples that were not prechlorinated had THM levels in the range of 35 to 45 μg/L. For the other sample, it is suspected that it was a case of some random contamination because all samples passed the QC criteria and a careful postmortem failed to show any irregularities. The preformed-chloramine train and in situ chloramine train showed quite close THM concentrations during this series of treatments.

For the 20-day test, of the pretreated samples, only the prechlorinated sample had a substantial THM level (about 70 μg/L, and this level was slightly higher than that observed for the 8-day test (about 65 μg/L). The other two were below 10 μg/L. Once again, the intermediate treatment THM levels were lower than expected for the prechlorination train (sample Nos. 14 and 15). As with the 8-day samples, there was no obvious difference between the different UV and Cl_2 sequences. Post-treatment THM levels showed a slight increase for the prechlorine train. The final concentrations seem to have increased by as much as 10 μg/L for the prechloramine train. All of the final treated samples that were not prechlorinated had THM levels in the range of 30 to 40 μg/L. The Cl_2/UV/PF CLM and Cl_2/UV/IS CLM sequences showed the advantage to the UV/Cl_2/IS CLM sequence (Nos. 24 and 25 versus 26; Nos. 27 and 28 versus No. 29).

THMs – DBP Test 5, Figure 5.13. For the 8-day test, of the pretreated samples, only the prechlorinated sample had a high THM level, which was approximately 90 μg/L. The other two pretreated samples were approximately 7 μg/L. These results are qualitatively similar to those in DBP Test Nos. 1 through 4.

The intermediate treatment THM levels were slightly lower than expected for the prechlorination train (sample Nos. 14 and 15). Sample Nos. 16 and 18 exhibited similar THM levels to sample Nos. 17 and 19. This indicates no obvious difference in the treatment sequence of chlorine and UV. Post-treatment THM levels reflect the intermediate levels. The final THM levels of the prechlorination train samples were slightly higher than those of the intermediate THM, which showed that postchloramination in the prechlorination train added little THMs. The final treated samples that were not prechlorinated had THM levels in the range of 40 to 50 μg/L. The preformed-chloramine train and in situ chloramine train showed nearly identical THM concentrations during this series of treatments.

For the 20-day test, of the pretreated samples, only the prechlorinated sample had a substantial THM level. The other two samples were below 10 μg/L. Once again, the intermediate treatment THM levels were lower than expected for the prechlorination train (sample Nos. 14 and 15). As with the 8-day samples, there was no obvious difference between the different UV and Cl_2 sequences. Post-treatment THM levels showed a slight increase for the prechlorine train. The final concentrations seem to have increased by as much as 10 μg/L for the prechloramine train. All of the final treated samples that were not prechlorinated had THM levels in the range of 35 to 45 μg/L.

HAA Tests

Figures 5.14 through 5.18 present all the HAA results for DBP Test Nos. 1 through 5.

HAAs – DBP Test No.1, Figure 5.14. For the 8-day test, the raw water (sample No. 1) had almost no HAAs. This was consistent with all the tests. Of the pretreated samples, only the prechlorinated sample had a substantial HAA level. The other two samples were below 15 μg/L, with the preformed-chloramine sample slightly lower than the in situ chloramine sample. The intermediate treatment sum of 5 haloacetic acids (HAA5) levels were elevated over those for the pretreated samples (sample Nos. 5 through 10). There was no reduction of HAA5 after UV treatment, but this is expected because the HAAs are not volatile. All of these paired samples seem to suggest that the sequence of intermediate chlorination followed by UV produces slightly lower HAA levels than the reverse. Most of this subtle effect seems to be lost upon final chloramination (sample Nos. 11 through 19). All of the final treated samples that were not prechlorinated had HAA5 levels in the range of 40 to 45 μg/L.

For the 20-day test, of the pretreated samples, only the prechlorinated sample had a substantial HAA level. In this case, it was approximately 80 μg/L rather than approximately 105 μg/L. The reasons for a reduction in HAA5 for a longer chlorine contact time are not clear. Biodegradation remains a possibility, but this seems unlikely due to the presence of a chlorine residual. The other two samples were below 20 μg/L, but this time the preformed-chloramine sample was higher than the in situ chloramine sample. The intermediate treatment HAA5 levels were equal to (sample Nos. 5 and 6) or higher than (sample Nos. 7 through 10) those for the pretreated samples. All of the final treated samples that were not prechlorinated had HAA5 levels in the range of 45 to 55 μg/L, which was slightly higher than that observed for the 8-day samples. In this case, it was quite clear that additional HAA formation occurred after final chloramination.

HAAs – DBP Test No. 2, Figure 5.15. For the 8-day test, of the pretreated samples, only the prechlorinated sample had a substantial HAA level. The other two samples were below 15 μg/L, with the preformed-chloramine sample slightly lower than the in situ chloramine sample. The intermediate treatment HAA5 levels were elevated over those for the pretreated samples (sample Nos. 14 through 19). There was no reduction of HAA5 after UV treatment, but this is expected because the HAAs are not volatile. Two of the three paired samples seem to suggest that the sequence of intermediate chlorination followed by UV produces slightly lower HAA levels than the reverse. Most of this subtle effect seems to be lost upon final chloramination (sample Nos. 21 through 29). All of the final treated samples that were not prechlorinated had HAA5 levels in the range of 30 to 45 μg/L.

For the 20-day test, of the pretreated samples, only the prechlorinated sample had a substantial HAA level. In this case, it was approximately 50 μg/L rather than approximately 45 μg/L in the 8-day test. The drop in DCAA observed during the 20-day test for Test No. 1 was not repeated here. The other two pretreatment samples were below 20 μg/L, but this time the preformed-chloramine sample was higher than the in situ chloramine sample. The intermediate treatment HAA5 levels were about equal to (sample Nos. 14 and 15) or higher than (sample Nos. 16 through 19) those for the pretreated samples. Except for one outlier that had an excessive monochloroacetic acid level (No. 26), all of the final treated samples that were not prechlorinated had HAA5 levels in the range of 30 to 50 μg/L, which was slightly higher than those observed for the 8-day samples. In this case, it was quite clear that additional HAA formation occurred after final chloramination. Most of this increase is due to increases in DCAA, a known monochloramine byproduct.

HAAs – DBP Test No. 3, Figure 5.16. For the 8-day test, of the pretreated samples, only the prechlorinated sample had a substantial HAA level. The other two samples were below 10 μg/L, with the preformed-chloramine sample slightly higher than the in situ chloramine sample. The intermediate treatment HAA5 levels of the chloramine trains were elevated over those for the pretreated samples (sample Nos. 16 through 19). There was no reduction of HAA5 after UV treatment. Two of the three paired samples seem to suggest that the sequence of intermediate chlorination followed by UV produces slightly lower HAA levels than the reverse. All of the final treated samples that were not prechlorinated had HAA5 levels in the range of 30 to 45 μg/L.

For the 20-day test, of the pretreated samples, only the prechlorinated sample had a substantial HAA level, and this level was nearly the same as that observed for the 8-day test (approximately 135 μg/L). The other two pretreatment samples were below 10 μg/L. The preformed-chloramine sample was higher than the in situ chloramine sample. The intermediate treatment HAA5 levels were about equal to (sample Nos. 14 and 15) or higher than (sample Nos. 16 through 19) those for the pretreated samples. All of the final treated samples that were not prechlorinated had HAA5 levels in the range of 35 to 50 μg/L, which was slightly higher than that observed for the 8-day samples.

HAAs – DBP Test No. 4, Figure 5.17. For the 8-day test, of the pretreated samples, only the prechlorinated sample had a substantial HAA level. The other two were well below 10 μg/L, with the preformed-chloramine sample slightly higher than the in situ chloramine sample. The intermediate treatment HAA5 levels of the chloramine trains were elevated over those for the pretreated samples (sample Nos. 16 through 19). There was no reduction of HAA5 after UV treatment. The three paired samples suggest that the sequence of intermediate chlorination followed by UV produces very nearly the same HAA levels as the reverse. All but one of the final treated samples that were not prechlorinated had HAA5 levels in the range of 30 to 45 μg/L. There was no obvious reason why sample No. 27 had a higher HAA5 reading.

For the 20-day test, of the pretreated samples, only the prechlorinated one had a substantial HAA level. The other two pretreatment samples were below 10 μg/L. The preformed-chloramine sample was very close to the in situ chloramine sample. The intermediate treatment HAA5 levels were about equal to (sample Nos. 14 and 15) or higher than (sample Nos. 16 through 19) those for the pretreated samples. All of the final treated samples that were not prechlorinated had HAA5 levels in the range of 25 to 35 μg/L.

HAAs – DBP Test No. 5, Figure 5.18. For the 8-day test, of the pretreated samples, only the prechlorinated sample had a substantial HAA level. The other two were well below 10 μg/L, with the preformed-chloramine sample slightly higher than the in situ chloramine sample. The intermediate treatment HAA5 levels of the chloramine trains were elevated over those for the pretreated samples (sample Nos. 16 through 19). There was no reduction of HAA5 after UV treatment. The three paired samples suggest that the sequence of intermediate chlorination followed by UV produces very nearly the same HAA levels as the reverse. The final treated samples that were not prechlorinated had HAA5 levels in the range of 40 to 55 μg/L.

For the 20-day test, of the pretreated samples, only the prechlorinated sample had a substantial HAA level. The other two pretreatment samples were below 10 μg/L. The preformed-chloramine sample was very close to the in situ chloramine sample. The intermediate treatment HAA5 levels were about equal to (sample Nos. 14 and 15) or higher than (sample Nos. 16 through 19) those for the pretreated samples. All of the final treated samples that were not prechlorinated had nearly identical HAA5 levels in the range of 40 to 45 μg/L.

Aldehydes

For the purpose of this report, total aldehydes are the sum of the formaldehyde, acetaldehyde, glyoxal, and methylglyoxal. These four compounds normally constitute the major low molecular weight aldehyde species. They are general oxidation byproducts, they are highly biodegradable, and their concentrations are thought to correlate with overall rapid biodegradability. Aldehyde levels for DBP Test Nos. 1 to 5 are summarized in Figures 5.19 to 5.23.

Aldehydes – DBP Test No. 1, Figure 5.19. The 8-day tests indicate that all treatments resulted in elevated aldehyde levels. However, pre- and intermediate chlorination seem to cause the greatest increase. The 20-day tests show that the final aldehyde levels (sample Nos. 21 through 29) are similar to those for the 8-day tests. However, pre- and intermediate chlorination seem to cause the greatest increase. Here, the relative effects of intermediate chlorination and postchloramination are clearer. Final chloramination quite consistently elevates the aldehyde levels up to near 40 μg/L.

Aldehydes – DBP Test No. 2, Figure 5.20. The 8-day tests show that all treatments resulted in elevated aldehyde levels. However, pre- and intermediate chlorination seem to cause the greatest increase. This is supported by results from DBP Test No. 1. When considering the final treated waters, there was a more even distribution across treatments, with the prechlorinated samples roughly equaling the in situ chloramined levels. The 20-day tests show final aldehyde levels that are similar, but not identical, to those for the 8-day pretreatment tests. The most striking treatment effects are for those of sample Nos. 16 versus No. 17 and No. 18 versus No. 19. These sets seem to indicate that the intermediate sequence UV/Cl_2 is far superior to Cl_2/UV in minimizing the formation of biodegradables. However, as indicated with other DBPs, this advantage is almost gone when viewing the final chloramined samples (Nos. 24 through 29). Final chloramination resulted in consistent aldehyde values in the range of 50 to 65 μg/L.

Aldehydes – DBP Test No. 3, Figure 5.21. The 8-day tests showed that all treatments resulted in elevated aldehyde levels. However pre- and intermediate chlorination seemed to cause the greatest increase. The postchloramination treatment did not add substantially to the aldehyde concentrations. When considering the final treated waters, the prechlorinated samples were approximately 10 μg/L higher than the prechloraminated samples. The 20-day tests show that 20-day aldehyde levels are slightly higher than those for the 8-day pretreatment tests. Again, the pre- and intermediate chlorination seem to cause the greatest increase in the aldehyde concentrations. The intermediate sequence UV/Cl_2 and Cl_2/UV showed similar effects on aldehyde formation. All of the final treated samples that were not prechlorinated had aldehyde levels in the range of 25 to 45 μg/L. The prechlorinated samples had final aldehyde levels in the range of 30 to 55 μg/L.

Aldehydes – DBP Test No. 4, Figure 5.22. The 8-day tests show that all treatments resulted in elevated aldehyde levels. All of the final treated samples that were not prechlorinated had aldehyde levels in the range of 30 to 40 μg/L. The prechlorinated samples had final aldehyde levels in the range of 30 to 50 μg/L. The 20-day tests show that aldehyde levels are slightly higher than those for the 8-day pretreatment tests. Again, the pre- and intermediate chlorination seemed to cause the greatest increase in the aldehyde concentrations. The intermediate sequence UV/Cl_2 and Cl_2/UV showed similar effects on aldehyde formation.

Aldehydes – DBP Test No. 5, Figure 5.23. The 8-day tests show that all of the final treated samples had aldehyde levels in the range of 30 to 40 μg/L. The 20-day tests show that aldehyde levels are a bit higher than those for the 8-day pretreatment tests. Again, the pre- and intermediate chlorination seemed to cause the greatest increase in the aldehyde concentrations. The intermediate sequence UV/Cl_2 and Cl_2/UV showed similar effects on aldehyde formation.

Ketoacids

The ketoacids are also general oxidation byproducts and indicators of biodegradable organic matter. In this report, the total ketoacids include glyoxalic acid and pyruvic acid. These compounds are more acidic and polar than the simple aldehydes. There are also indications that they are more rapidly biodegradable and possibly more reactive with chlorine-based disinfectants (Xie and Reckhow, 1992; 1997; Garcia-Araya et al., 1995; Schmidt et al., 1999). Ketoacid results for DBP Test Nos. 1 through 5 are summarized in Figures 5.24 to 5.28.

Ketoacids – DBP Test No. 1, Figure 5.24. The 8-day tests show that the ketoacids are formed in greater concentrations than the aldehydes. Again, free chlorination seems to produce the largest amount of these byproducts. However, postchloramination resulted in a substantial drop in the concentration of the ketoacids. This may be due to direct reactions between monochloramine and this group of compounds. The final concentration of ketoacids was only 50 to 100 percent higher than the level in the raw water.

The 20-day tests show that the ketoacids are reduced following postchloramination. The final concentration is quite similar across the treatments. The application of free chlorination does not seem to have affected the postchloramination levels.

Ketoacids – DBP Test No. 2, Figure 5.25. The 8-day tests show that with this water, the ketoacids were formed in slightly lower concentrations than the aldehydes. Again, free chlorination seems to produce the largest amount of these byproducts, but the differences are minor. In DBP Test No.1, postchloramination resulted in a substantial drop in the concentration of the ketoacids, but this was not seen in DBP Test No. 2. The final concentration of ketoacids was approximately 40 μg/L with relatively minor differences attributable to differences in treatment.

The 20-day tests show that the ketoacids are reduced following postchloramination. The final concentration is quite similar across the treatments. Surprisingly prechlorination did not result in elevated levels, especially when viewing the final concentrations.

Ketoacids – DBP Test No. 3, Figure 5.26. The 8-day tests show that with this water, the ketoacids were formed in slightly lower concentrations than the aldehydes. Again, free chlorination seems to produce the largest amount of these byproducts. In this test, postchloramination resulted in a substantial drop in the concentration of the ketoacids. This may be due to the possible biodegradation during the 5-day post-treatment. The final concentration of ketoacids was approximately 10 to 18 μg/L with relatively minor differences attributable to differences in treatment.

The 20-day tests show that the ketoacids are reduced following postchloramination. The final concentration is quite similar across the treatments. Surprisingly, prechlorination did not result in elevated levels from the perspective of the final treated samples.

Ketoacids – DBP Test No. 4, Figure 5.27. The 8-day tests show that with this water, the ketoacids were formed in slightly lower concentrations than the aldehydes. Again, free chlorination seems to produce the largest amount of these byproducts. The final concentration of ketoacids was about 15 to 35 μg/L. In general, final concentrations in post-treatment were lower than those observed after intermediate treatment. This had been noted previously and may be attributable to biodegradation during the 5-day post-treatment period.

The 20-day tests show that free chlorination produces the largest amount of these byproducts when considering the pretreated samples. In this test, postchloramination resulted in a substantial drop in the concentration of the ketoacids. Again, this may be due to possible biodegradation. The final concentration was quite similar across the treatments. Surprisingly, prechlorination did not result in elevated levels when viewing the final treated samples.

Ketoacids – DBP Test No. 5, Figure 5.28. The 8-day tests show that with this water, the ketoacids were formed in slightly lower concentrations than the aldehydes. Again, free chlorination produces the largest amount of these byproducts. The final ketoacid levels were only slightly higher than those of the intermediate ketoacids, which showed that postchloramination added little or no ketoacids. The final concentration of ketoacids was approximately 20 to 30 μg/L for all samples. This indicates that the advantage at first exhibited by the prechloramination train was lost upon subsequent treatment.

The 20-day tests show that free chlorination produces the largest amount of the ketoacids when considering the pretreated samples. In this test, postchloramination resulted in a slight increase in the concentration of the ketoacids. The final concentration was quite similar across all treatments.

Testing for Biodegradable Organic Matter

For DBP Test No. 4, BDOC and AOC were measured for two selected trains following the 8-day pretreatment protocol. One of the selected trains was in situ chloramination followed by chlorination and UV, and then post-in situ chloramination (IS/Cl_2/UV/IS). The other was in situ chloramination followed by UV and chlorination, and then post-in situ chloramination (IS/UV/Cl_2/IS). The results are shown in Table 5.3. These two treatment trains showed nearly identical AOC and BDOC concentrations.

Both the BDOC and AOC concentrations are rather high, suggesting biologically unstable water. The levels measured are probably commensurate with the high TOC and the long-term maintenance of a chlorine/chloramine residual.

DBP Test Nos. 1 Through 5 Conclusions

THMs

TTHM levels, as produced by the various disinfection scenarios, resulted in concentrations between 25 and 55 μg/L for prechloramination scenarios and 45 to 120 μg/L for those involving prechlorination. Actual THM values expected in the Winnipeg system may be different due to different levels of volatilization in the Deacon Reservoir. THM results and trends were consistent for DBP Test Nos. 1 through 5. THM levels did not change substantially after chloramination. Minor changes in THMs were observed depending on the sequence of intermediate chlorination and UV treatment. However, these are not deemed significant due to the uncertainty of THM volatilization during bench-scale UV treatment.

Table 5.3
Results of BOM assays

Treatment train	AOC (μg/L)	BDOC (mg/L)
IS/Cl_2/UV/IS	513	1.17
IS/UV/Cl_2/IS	533	1.36

HAA5

The haloacetic acids were found in the final treated samples across a range of 25 to 55 μg/L when prechloramination was used. The corresponding values for prechlorination trains were 40 to 100 μg/L. Some biodegradation of DCAA might have occurred in pretreatment, but this compound also showed long-term formation over the 5-day contact with chloramines No. 1 through 5. A small, but consistent increase in HAA5 was noted after intermediate treatment when UV preceded chlorination as compared to the reverse order. A similar consistent difference was not observed when looking at the final postchloraminated samples. Postchloramination substantially increased the concentration of HAAs.

Aldehydes and Ketoacids

Aldehyde data showed a substantial increase due to the application of chlorine-based disinfectants. Final aldehyde levels were in the range of 25 to 55 μg/L.

However, ketoacid levels were only moderately elevated. Final ketoacid levels were in the range of 10 to 40 μg/L. There seemed to be very little difference in the amount of nonchlorinated byproducts formed when comparing chlorination versus chloramination.

Process Sequencing

The DBP data showed some consistent qualitative trends that bear on the issue of process selection. First, there was no obvious difference between the UV/Cl_2 and Cl_2/UV sequences. Although some tests showed the sequence of Cl_2/UV to offer a modest advantage in controlling THM and HAA concentrations, other tests showed the reverse. Furthermore, the benefits in using one sequence versus the other were usually dampened or completely lost when viewing the final chloraminated data. Bioassay data (AOC and BDOC) also indicate that the sequencing of chlorine and UV is not important. Secondly, there was no clear indication that preformed chloramines led to reduced levels of DBPs as compared to in situ formed chloramines.

DBP TEST NOs. 6, 7, AND 8

Modification to Disinfection Byproduct Studies

On the basis of consistent and similar results of DBP Test Nos. 1 through 5, the DBP test program was modified to investigate other areas of interest with respect to disinfection and DBPs that may have potential impacts on Winnipeg's system. The modified DBP Test Nos. 6, 7, and 8 and objectives are summarized as follows:

- DBP Test No. 6 – UV impact on free-chlorine demand and disinfection byproduct formation.
  - To determine if UV causes an increase in free-chlorine demand and how much additional chlorine would have to be added.
  - If there is an increase in demand, determine if there are any detectable effects on DBP formation of this phenomenon.
- DBP Test No. 7 – Kinetics of disinfection byproduct formation from free chlorination of Shoal Lake Water.
  - To investigate various disinfection scenarios.
  - If free chlorination continued at headworks, test under different conditions, dose, temp, pH, etc.
- DBP Test No. 8 – UV impact on free-chlorine demand and emerging disinfection byproducts.
  - To determine potential NDMA levels in Shoal Lake and finished water.
  - To determine if the sequence of chlorination and UV might have any effect.

This section discusses test protocols and presents results from DBP Test Nos. 6, 7, and 8.

DBP Test No. 6 – UV Impact on Free-Chlorine Demand and Disinfection Byproduct Formation

Introduction

In some of the "intermediate testing" for DBP Test Nos. 1 through 5, it was observed that the FRC was (0.05 to 0.15 mg/L) lower when the water was pretreated with UV (i.e., UV/Cl_2 sequence versus Cl_2/UV). This was unexpected and warranted further investigation to clarify the UV effect on FRC. In addition, if UV causes an increase in free-chlorine demand, is there any detectable effect on the formation of DBPs? The test plan DBP Test No. 6 was designed to study the impact of UV treatment on free-chlorine demand and selected DBPs.

Bench-Scale Testing

As with DBP Test Nos. 1 through 5, bench-scale testing for DBP formation was conducted in the UMass laboratories. Raw water samples of approximately 40 L were collected, and samples of approximately 14 L were treated in a series of batch and flow-through treatment systems aimed at simulating the various pre- and postdisinfection steps. The preoxidation step involved the addition of ammonia followed by chlorine (a more conventional chloramination practice). Two possible intermediate treatment scenarios were applied after pre-treatment as follows:

- Free chlorination for virus control followed by UV followed by application of ammonia to form chloramines
- UV treatment followed by a short free-chlorine contact and then postammoniation

Results

Raw Water Quality. Table 5.4 presents Shoal Lake raw water quality data for the day in which the bulk water sample was collected.

These data indicate a SUVA of 2.08 L/mg/m. This is higher than the raw water used in DBP Test Nos. 1 through 5. This indicates that the water has a higher humic character than the former samples; however, it is still low for water of this TOC level. The TOC for this series of tests was higher than for Test No. 2 and lower than Nos. 1 and 3 through 5.

Disinfectant Dose/Residual Tests. Based on disinfectant demand tests, disinfectant doses were selected to achieve the desired residual after 30 hours.

The key numbers (No. 1 to No. 46) represented in Figure 5.29 correspond to the points for which chlorine data are available in the test. At each important experimental juncture, chlorine residuals were measured. This includes the measurement of free chlorine, monoCLM, and diCLM.

Forty-one of these sample points represent actual chlorine residual measurements. However, five of these (Nos. 2, 4, 5, 6, and 28) are reserved for dosed values, which should reflect the concentrations at the instant of chlorine addition. The dosed values correspond to:

- Pretreatment dose (No. 2)
- Intermediate treatment doses: $Cl_2(1)$ = 1.5 mg Cl^2/L (No. 4); $Cl_2(2)$ = 2.5 mg Cl_2/L (No. 5); $Cl_2(3)$ = 3.5 mg Cl_2/L (No. 6)
- Post-treatment dose: to achieve the desired residual at the end of the 5-day post-treatment period (No. 28)
- The UV doses used: UV(1) = 0 mJ/cm^2; UV(2)=400 mJ/cm^2; UV(3) = 2000 mJ/cm^2

Table 5.4
Raw water quality – DBP Test No. 6

DBP test no.	Date	Temp °C	pH	UV 254 cm-1	TOC (mg/L)	NH3-N (mg/L)
6	18-Mar-02	2.0	7.1	0.166	7.98	0.09

Chlorine doses and residuals are depicted in Figure 5.30. The data indicate that the chlorine residual was essentially depleted at the end of the 8-day pretreatment period (sample No. 3). This was intended because the preoxidant dosing criteria called for a small (0.2 mg/L) residual after only 30 hours.

Among the samples treated with UV after chlorine (Nos. 10 through 18), only the controls (i.e., 0 mJ/cm^2 UV dose; Nos. 10, 13, and 16) and one experimental (high-chlorine dose sample [No. 17]) had measurable FRC (0.1 to 0.4 mg/L). This reflects the ability of UV treatment to destroy FRC. Furthermore, for the samples treated with chlorine after UV (Nos. 19 through 27), the FRC decreased with increasing UV dose as UV appeared to create some free-chlorine demand (No. 19 versus No. 22 versus No. 25; No. 20 versus No. 23 versus No. 26; No. 21 versus No. 24 versus No. 27). This confirms that the higher the UV dose, the higher the chlorine demand.

The residuals after post-treatment indicate that the final chloramine dose (1.1 mg/L) was near optimal because most of the final samples had combined residuals around the 0.6 mg/L target value.

Disinfection Byproducts

THMs. The sampling key for DBPs is exactly the same as for the chlorine doses and residuals (Figure 5.29). Figure 5.31 presents all of the THM data. As expected, the raw water (sample No. 1) had almost no THMs. The THM level after the 8-day in situ chloramination is approximately 10 µg/L.

The intermediate treatment results indicate an increase in THMs with the addition of free chlorine. The results of the samples treated with chlorine after UV (Nos. 19 through 27) failed to show a convincing trend between THM levels and UV dose (No. 19 versus No. 22 versus No. 25; No. 20 versus No. 23 versus No. 26; No. 21 versus No. 24 versus No. 27).

For post-treatment, THM levels reflect the intermediate levels. This is not unexpected because postchloramination usually does not add much to THM concentrations.

HAAs. Figure 5.32 contains all of the HAA data for the pre-treatment and intermediate treatment tests. Similar to the THM result, the raw water (sample No. 1) had almost no HAAs. The HAA5 level is approximately 5 μg/L after 8 days of pre-treatment. Similar to the THMs results, the HAA results also show a substantial increase in HAAs with the addition of free chlorine. In this case, it is quite clear that the higher the UV dose, the higher the HAAs. Because the chlorine dose was held constant, this increase in HAA formation (and increase in chlorine demand) is attributed to some change in the organic precursors induced by UV irradiation. Also, it is evident that the UV/Cl_2 sequence produces lesser amounts of HAAs than the corresponding Cl_2/UV sequence (Nos. 19 through 27 versus Nos. 10 through 18).

Post-treatment HAA levels reflect the intermediate levels. This is not unexpected, as postchloramination usually does not add much to HAA concentrations.

Raw Water Tests

The raw water tests were done to better isolate the effect of UV treatment on free-chlorine demand and DBP formation. Approximately 6 L of raw water was subjected to the UV/Cl_2 sequence. Three batches each of 2 L were treated with three UV doses, namely 0 mJ/cm^2; 400 mJ/cm^2, and 2,000 mJ/cm^2. Each batch was then subjected to six free-chlorine doses (2, 2.5, 3, 3.5, 4, and 5 mg Cl_2/L). After a 20-minute incubation at 2°C, samples were taken to determine the FRC and DBP concentrations. Figure 5.33 shows the chlorine residual data.

It appears that when UV dose was increased from 0 mJ/cm^2 to 400 mJ/cm^2, there was a slight indication of increased free-chlorine demand. When the UV dose was increased to 2,000 mJ/cm^2, there was a small increase in free-chlorine demand. With only 1 replicate, it is difficult to draw strong conclusions. Figure 5.34 shows the THM and HAA5 concentrations from this same experiment. It is clear that THM concentrations increase with the increase of UV dose, especially for the 2,000 mJ/cm^2 dose. The same trend was observed for HAA data.

Conclusions

The following conclusions are based on DBP Test No. 6 as summarized in this section.

UV treatment caused a small (i.e., 10 percent) increase in chlorine demand. The increase in demand was greater at higher UV doses.

In general, DBPs (mostly HAAs) also increased with increasing UV dose. In some cases, these increases were as high as 20 percent over the control.

This test showed a trend indicating lower DBPs formation in the UV/Cl_2 sequence as compared to the reverse (Cl_2/UV).

DBP Test No. 7 – Kinetics of Disinfection Byproduct Formation From Free Chlorination of Shoal Lake Water

Introduction

In this set of experiments, a data set was developed to support a partial DBP formation model. This model might be used as a stand-alone tool for assessing water quality and process changes or as a water quality add-on to a hydraulic model. This work involved the collection of a 40-L water sample from Shoal Lake. The sample was subject to a matrix of chlorination tests, whereby the following conditions were varied:

- pH
- Chlorine dose
- Contact time

The bulk water sample was subjected to an orthogonal matrix of five experiments. These experiments were designated as presented in Table 5.5.

Table 5.5
Orthogonal matrix for DBP Test No. 7

Experiment no.	Description
1	Baseline (ambient)
2	High-chlorine dose (pH=ambient)
3	Low-chlorine dose(pH=ambient)
4	High pH (pH=9)
5	Low pH (pH=6)

The values for the high and low conditions depend on the observed range in the system. The chlorination conditions are presented in Table 5.6.

High-chlorine doses were set at an additional 100 percent of the baseline dose, and low-chlorine doses were set at a reduced value of 30 percent than the ambient dose. High pH was set at 9 using sodium hydroxide (NaOH), and low pH was set at 6 using dilute sulfuric acid (H_2SO_4). The samples were split into BOD bottles where they were chlorinated, adjusted to the desired pH, and held for contact times of 0.5, 2, 8, 24, 72, and 168 hours. After each contact time, one BOD bottle was opened and tested for chlorine demand and pH. Samples for THM and HAA were also taken.

Results

Raw Water Quality. Table 5.7 presents the raw water quality data for the day in which the bulk water sample was collected.

Table 5.6
Chlorination conditions

Parameter	Baseline conditions	High/low values
pH	Ambient	6, 9
Chlorine dose	Level used for prechlorination	-30%, +100%
Contact time	0.5, 2, 8, 24, 72, 168 hr.	

Table 5.7
Raw water quality – DBP Test No. 7

DBP test no.	Date	Temp °C	pH	UV 254 cm-1	TOC (mg/L)	NH3-N (mg/L)
7	5-June-02	16.5	7.48	0.141	7.56	NA

These data indicate an SUVA of 1.87 L/mg/m. The SUVA for this water falls in the intermediate range when compared to the previous samples. However, it is still low for water of this TOC level.

Based on the protocol for this run, the matrix of experimental conditions is presented in Table 5.8.

Chlorine Residuals. Figure 5.35, Figure 5.36, and Figure 5.37 show the chlorine residuals for the varying conditions of pH for each chlorine dose. The chlorine residual decreases rapidly, for all the cases, within the first 2 hours of dosing. The chlorine residual approached zero for the chlorine dose of 3.4 mg/L (Figure 5.35) at approximately 24 hours. For the chlorine dose of 4.8 mg/L (Figure 5.36), the residual approached zero at about 72 hours. The chlorine residual does not reach zero (Figure 5.37) even at 168 hours for the chlorine dose of 9.6 mg/L.

It is evident from the data that the residual chlorine is slightly lower for the lower pH experiments, within the first 10 hours of dosing. Beyond the 10-hour time frame, this pattern does not exist. For the chlorine doses of 4.8 mg/L and 9.6 mg/L, the above pattern is reversed after 10 hours.

THM Results. Figure 5.38, Figure 5.39, and Figure 5.40 contain all of the THM data for the three chlorine doses under varying conditions of pH. For each specific pH and contact time, the THM concentration increased with chlorine dose, as expected. The highest THM concentrations were 90 μg/L, 130 μg/L, and 240 μg/L for chlorine doses of 3.4 mg/L, 4.8 mg/L, and 9.6 mg/L, respectively. For all three doses, the THM concentration increased rapidly for the first 24 hours. For the chlorine dose of 3.4 mg/L, there was no significant increase in THM concentration beyond 24 hours. The same trend was observed with the chlorine dose of 4.8 mg/L after 72 hours. For the chlorine dose of 9.6 mg/L, the THM concentration increased at a comparatively higher rate throughout the duration of the experiment.

Table 5.8
Experimental matrix for DBP Test No. 7

Parameter	Baseline conditions	High values	Low values
pH	7.48	9	6
Chlorine dose (mg/L)	4.8	9.6	3.4
Contact time (hr.)	0.5, 2, 8, 24, 72, 168		

From the three figures, it is clear that the THM formation increases with increasing pH, at least up to pH 9, regardless of the chlorine dose or reaction time.

HAA5 Results. Figure 5.41, Figure 5.42, and Figure 5.43 contain all of the HAA5 data for the three chlorine doses at varying conditions of pH. For each specific pH and contact time, the HAA5 concentration increased with chlorine dose, as expected. The highest HAA concentrations were 77 μg/L, 125 μg/L, and 250 μg/L for chlorine doses of 3.4 mg/L, 4.8 mg/L, and 9.6 mg/L, respectively. Similar to THM formation, the HAA5 concentration also increased in a monotonic fashion with time. For all three doses, the HAA5 concentration increased rapidly for the first 24 hours. For the chlorine dose of 3.4 mg/L, there was no significant increase in HAA5 concentration beyond 24 hours. The same trend was observed with the chlorine dose of 4.8 mg/L after 72 hours. For the chlorine dose of 9.6 mg/L, the HAA5 concentration increased at a comparatively higher rate throughout the duration of the experiment.

From the three figures, it is clear that the HAA5 formation diminishes with increasing pH under alkaline conditions, which is the exact opposite trend to that observed with THM. This has generally been observed by others who have studied HAA formation as a function of pH.

DBP Modeling. DBP formation depends on many parameters, such as chlorine dose, pH, contact time, temperature, TOC, UV254, etc. During the past two decades, various empirical models have been developed for DBP formation, especially the power function model (Amy et al., 1987; 1993; Chowdhury et al., 1994). The general form of this equation is presented below in Equation 5.1:

$$DBP = a\,(TOC)^{b}\,(UV)^{c}\,(1/DOC)^{d}\,(Cl2)^{e}\,(t)^{f}\,(T)^{g}\,(pH+h)^{i}\,(Br+j)^{k} \qquad (5.1)$$

For this series of experiments, three major parameters, namely, chlorine dose, contact time, and pH, were studied regarding DBP formation to develop a partial DBP formation model. The general form of the model is given by Equation 5.2 (below) (Amy et al., 1987; 1993; Chowdhury et al., 1994). A multiple linear regression method was used to determine the parameters associated with the model. For the multiple linear regression, the log transformed DBP data were the dependent variables, and all other log transformed parameters such as chlorine dose, pH, and time were independent variables.

$$\log(DBP) = a + b\log(\text{time [hrs.]}) + c\log(\text{chlorine dose [mg/L]}) + d\log(pH) \qquad (5.2)$$

The antilog of Equation 5.2 allows for the direct calculation of the DBP formation. This is given by Equation 5.3:

$$DBP = 10a\,(\text{Time [hrs]}b\,(Cl^{2}\text{ dose [mg/L]})c\,(pH)d \qquad (5.3)$$

For these tests, the best-fit coefficients for Equation 5.3 are presented in Table 5.9.

The statistical parameter showing the accuracy of prediction by the model for each DBP, namely R^2, and the standard error of estimate for the DBP by the model is presented in Table 5.10.

Table 5.9
Values for power function model for DBP formation

Model	a	b (time)	c (Cl_2)	d (pH)
THM	-0.021	0.191	0.444	1.518
$CHCl_3$	-0.095	0.195	0.450	1.564
HAA5	1.771	0.128	0.718	-0.578
DCAA	0.504	0.126	0.488	0.673
TCAA	2.788	0.183	0.899	-2.420

Table 5.10
Statistical values from linear regression analysis

Model	R	R^2	Adjusted R^2	Std. error of estimate
THM	0.965	0.932	0.928	0.061
CHCl3	0.967	0.934	0.930	0.061
HAA5	0.921	0.847	0.838	0.079
DCAA	0.939	0.882	0.875	0.057
TCAA	0.909	0.825	0.815	0.139

From this model, the coefficients "b" and "c," the power function of time and chlorine dose, respectively, are positive for the DBPs modeled. This means that increases in both chlorine dose and contact time will have an increasing effect on DBP formation. The values for the power function "d" of pH for THM and Chloroform formation are positive, which means that as the pH increases, the concentration of THM and chloroform will also increase. This is consistent with the experimental data. The values for "d" are negative for HAA and TCAA, implying that as the pH increases, the concentration of HAA and TCAA will decrease. For DCAA, positive values of "d" are observed (i.e., DCAA concentration will increase with pH). This is consistent with the experimental data.

Conclusions

The following conclusions are based on the DBP Test No. 7 as summarized in this section.

The partial DBP modeling for this run yielded the following equations for THMs and HAA5:

$$\text{THM } (\mu g/L) = 10^{-0.021} + (\text{Cl2 dose [mg/L]})^{0.444} + (\text{Time [hrs.]})^{0.191} + (\text{pH})^{1.518}$$
$$\text{HAA5 } (\mu g/L) = 10^{1.771} + (\text{Cl2 dose [mg/L]})^{0.718} + (\text{Time [hrs.]})^{0.128} + (\text{pH})^{-0.578}$$

DBP formation increases with chlorine dose and contact time under all pH conditions.

THM concentration increased with pH, while HAA5 concentration decreased with pH at all doses and contact times.

The general trends for the kinetic data for this water were generally in agreement with those reported for other waters in the literature.

DBP Test No. 8 – UV Impact on Free-Chlorine Demand and Emerging Disinfection Byproducts

Introduction

The Shoal Lake water is relatively unusual due to its relatively high TOC level but relatively low SUVA values. While HAA and THM formation kinetics from this water are known, detailed investigation of other byproducts was not undertaken. Specifically, the TOX represents the total amount of halogenated byproduct formation. While TOX is not currently regulated, the possibility of future regulation exists. NDMA is currently regulated in California and Ontario, and regulations could reach the federal level. This byproduct also seems to be formed in chloraminating systems, and it is susceptible to decomposition by UV light. If present, the sequence of chlorination and UV may be important.

The tests for this water were designed to study the impact of UV on free-chlorine demand and these emerging DBPs.

Bench-Scale Testing

Bench-scale testing for TOX and NDMA was conducted in the UMass laboratories. Raw water samples of approximately 40 L were collected and shipped to UMass for testing.

Samples of approximately 14 L were treated in a series of batch and flow-through treatment systems aimed at simulating various pre- and postdisinfection (see Figure 5.44). The preoxidation is ammonia followed by the addition of chlorine (a more conventional chloramination practice).

Oxidant/disinfectant doses are selected to achieve a measurable residual at the end of 30 hours. This is decided upon from a set of preliminary bench-scale demand tests before treating the full bulk sample.

Eight-days' contact time, representing the range that is likely to be employed in Winnipeg's system, is used for this work. At the start of the experiment and at the end of the contact time, samples are collected for analysis of oxidant residual and DBPs (locations No. 1 and No. 2 in Figure 5.44). The DBPs that are measured include the THMs (and other neutral-extractables) and the HAAs (all 9).

Two possible intermediate treatment scenarios (Figure 5.45) are applied after pretreatment, as follows:

- Free chlorination for virus control followed by UV and application of preformed chloramines
- UV treatment followed by a short free-chlorine contact and then postammoniation

Three chlorine doses and three UV doses were used. Combined, this resulted in 18 overall treatment permutations. Samples were collected immediately after intermediate treatment (Nos. 10 through 18 for Cl_2/UV and Nos. 19 through 27 for UV/Cl_2) and after post-treatment (Nos. 29 through 46). The final samples (Nos. 29 through 46) are collected after a 5-day contact time with the residual disinfectants. This is intended to simulate an extreme distribution system residence time. As with the pretreatment analyses, all samples are measured for residual disinfectant, as well as THMs and HAAs. Six samples, three with the same chlorine dose and varying UV dose and three with the same UV dose and varying chlorine dose, were tested for TOX. Two samples from a parallel treatment train with a different UV/Cl_2 sequence were tested for NDMA.

UV treatment was conducted at UMass in a 2-L flow-through reactor using a monochromatic lamp. The equipment was brought to UMass from UNH for use throughout the project, and Dr. Malley and his team trained the UMass research assistants in its use.

Results

Raw Water Quality. Table 5.11 presents the raw water quality data for the day in which the bulk water sample was collected.

These data indicate an SUVA of 1.68 L/mg/m. This is higher than Test Nos. 1 through 5, but lower than Test Nos. 6 and 7. This indicates that the water has a somewhat higher humic character than the first five waters. However, it is still quite low for water of this TOC level.

Preliminary Disinfectant Demand Tests

Chloramines. One of the first experiments that was run with the raw water was an oxidant demand study using in situ formed chloramine. A dose/residual curve was developed so that the appropriate dose to achieve the desired residual (0.2 mg Cl_2/L) at the end of the preoxidation contact time (30 hours) could be estimated. The resulting curve is depicted in Figure 5.46

From the demand test, the appropriate dose for bench-scale pretreatment was determined to be 0.43 mg/L.

Treated Water Chlorine Residuals. At each important experimental juncture, chlorine residuals were measured using the DPD method. This includes the measurement of free chlorine, monoCLM, and diCLM.

The key numbers (Nos. 1 to 46) are represented in Figure 5.47 and correspond to the different points for which chlorine data are available in the test.

Table 5.11
Raw water quality – DBP Test No. 8

DBP test no.	Date	Temp °C	pH	UV 254 cm-1	TOC (mg/L)	NH3-N (mg/L)
8	13-Aug-02	20.0	7.1	0.161	9.60	NA

Forty-one of these points represent actual chlorine residual measurements. However, five (Nos. 2, 4, 5, 6, and 28) are reserved for dosed values, which should reflect the concentrations at the instant of chlorine addition. These dosed values correspond to:

- Pretreatment dose (No. 2)
- Intermediate treatment doses: $Cl_2(1)$ = 1.5 mg Cl_2/L (No. 4); $Cl_2(2)$ = 2.5 mg Cl_2/L (No. 5); $Cl_2(3)$ = 3.5 mg Cl_2/L (No. 6)
- Post-treatment dose: to achieve the desired residual at the end of the 5-day post-treatment period (No. 28)
- The UV doses used: UV(1) = 0 mJ/cm^2; UV(2) = 400 mJ/cm^2; UV(3) = 2000 mJ/cm^2

Chlorine doses and residuals are presented in Figure 5.48. Figure 5.48 indicates that the chlorine residual was essentially depleted at the end of the 8-day pretreatment period (sample No. 3). This is to be expected as the preoxidant dosing criteria called for a small (0.2 mg/L) residual after only 30 hours. Among the samples treated with UV after chlorine

(Nos. 10 through 18), only the samples with 0 mJ/cm^2 UV dose (Nos. 10, 13, and 16) and one sample with higher chlorine dose (No. 17) had a measurable FRC This reflects the ability of UV treatment to destroy the FRC.

In the samples treated with chlorine after UV (Nos. 19 through 27), the FRC slightly decreased (approximately 10 percent at lower dose) with increasing UV dose (No. 19 versus No. 22 versus No. 25; No. 20 versus No. 23 versus No. 26; No. 21 versus No. 24 versus No. 27). In the case of the highest chlorine dose set, there also seems to be an overall drop in the total residual with increasing UV dose. From the chlorine residuals after post-treatment, it is evident that the final chloramine dose (1.1 mg/L) was near optimal because most of the final samples had combined residuals around the 0.6 mg/L target value.

Chlorinated Byproducts

The sampling key for the DBPs is exactly the same as for the chlorine doses and residuals (Figure 5.47).

THMs. Figure 5.49 shows all the THM concentrations for the pretreatment and inter-treatment. As expected, the raw water (sample No. 1) had almost no THMs. The THM level after 8-day in situ chloramination is approximately 13 μg/L. The intermediate treatment results indicate that with the addition of free chorine, the amount of THM increases substantially, and the higher the chlorine dose, the higher the THM level. The results of the samples treated with chlorine after UV (Nos. 19 through 27) showed a subtle trend of increasing THM levels with increasing UV dose (No. 21 versus No. 24 versus No. 27; No. 19 versus No. 22 versus No. 25). There is no consistent difference between the UV/Cl_2 train (Nos. 19 through 27) and Cl_2/UV train (Nos. 10 through 18) in controlling THM formation. Some dose combinations led to higher levels using the UV/Cl_2 sequence (No. 12 versus No. 25, No. 17 versus No. 24), and some were lower (No. 11 versus No. 22, No. 14 versus No. 23, No. 15 versus No. 26). There is too little information to provide insight into the mechanism. Post-treatment THM levels reflect some of the same trends as the intermediate treatment. Post-treatment THM concentrations show an increase of approximately 10 to 20 μg/L for samples with a higher chlorine dose. This may be due to the residual chlorine in these samples after inter-treatment.

HAAs. Figure 5.50 shows all the HAA results. Similar to the THM result, the raw water (sample No. 1) had approximately 2 μg/L HAAs. The HAA5 level is approximately 5 μg/L after 8-day pretreatment. Similar to the THMs results, the HAA results also show a substantial increase of HAAs with addition of free chlorine. In this case, it also shows a subtle trend whereby the higher the UV dose, the higher the HAAs.

As with the THMs, there is no obvious difference between the UV/Cl_2 train (Nos. 19 through 27) and Cl_2/UV train (Nos. 10 through 18) in controlling HAA formation. Post-treatment HAA levels reflect the intermediate levels. This is not unexpected, as postchloramination usually does not add much to HAA concentrations.

TOX and NDMA Results. Sample Nos. 32 through 34 and No. 41 through 43 were selected for TOX analysis. The results are shown in Table 5.13.

Table 5.12
TOX analysis results

Sample number	UV dose (mJ/cm2)	Cl2 dose (mg/L)	Sequence	TOX (µg Cl/L)
32	0	2.5	Cl2 / UV	106.5
33	400	2.5	Cl2 / UV	142.1
34	2,000	2.5	Cl2 / UV	172.9
41	400	1.5	UV / Cl2	104.7
42	400	2.5	UV / Cl2	204.8
43	400	3.5	UV / Cl2	242.8

For sample Nos. 32 through 34 with the same chlorine dose of 2.5 mg/L, TOX concentration increased with increasing UV dose. This increase was as high as 34 percent when UV dose went from 0 mJ/cm^2 to 400 mJ/cm^2 and 62 percent when the dose was increased to 2,000 mJ/cm^2. As expected, TOX increased with chlorine dose for the same UV dose of 400 mJ/cm^2 (sample Nos. 41 through 43).

For the samples with different UV and Cl_2 sequences (Nos. 33 and 42), the sample chlorinated after UV treatment (UV/Cl_2) shows higher TOX concentration than the sample chlorinated before UV treatment (Cl_2/UV). This increase in TOX concentration is approximately 44 percent (No. 42 versus No.33). It does not appear that destruction of the chlorine residual by UV was a factor, as both (No. 33 and 42) had about the same residual chlorine level after treatment. Rather it seems that UV treatment changed the NOM so that it was more reactive with chlorine.

Sample Nos. 33 and 42 were selected for NDMA analysis. The results are shown in Table 5.13.

From Table 5.12, the NDMA concentration of UV/Cl_2 sequence (No. 42) is almost doubled compared to the Cl_2 /UV sequence (No. 33). There is not enough information to provide insight on the mechanism.

Table 5.13
NDMA analysis results

Sample number	Sample identification	NDMA (ng/L)
33	IS/Cl_2(2)/UV(2)/PF	5.6
42	IS/UV(2)/Cl_2(2)/PF	11

Raw Water Tests. The raw water tests were done to better isolate the effect of UV treatment on subsequent free-chlorine demand and DBP formation. Approximately 6 L of raw water was subjected to the UV/Cl_2 sequence. Three batches each of 2 L were treated with three UV doses, 0 mJ/cm^2, 400 mJ/cm^2, and 2,000 mJ/cm^2, respectively. Each batch was then subjected to six free-chlorine doses (2, 2.5, 3, 3.5, 4, and 5 mg Cl_2/L). After a 20-minute incubation at 20°C, samples were taken to determine the FRC and DBP concentrations.

Figure 5.51 shows the chlorine residual data. Chlorine demand increased by 0.1 to 0.2 mg/L when the UV dose was increased from 0 mJ/cm^2 to 400 mJ/cm^2. An increase of 0.2 to 0.5 mg Cl_2/L was observed when the UV dose was increased from 0 mJ/cm^2 to

2,000 mJ/cm^2. It is clear that UV irradiation increases the chlorine demand in the subsequent chlorination process.

Figure 5.52 shows the THM concentrations from this same experiment. The THM concentrations for the UV dose of 400 mJ/cm^2 increased by up to 12 percent compared to the 0 mJ/cm^2 UV dose. This increase was as high as 40 percent in the case of the 2,000 mJ/cm^2 UV dose. It is clear that THM concentrations increase with the increase of the UV dose.

CONCLUSIONS

Following are conclusions based on DBP Test No. 8 as summarized in this section.

There was no obvious difference between the UV/Cl_2 and Cl_2/UV sequences with respect to the THM and HAA formation. However, at UV doses greater than 400 mJ/cm^2, this run showed a clear trend indicating lower TOX and NDMA formation in the Cl_2/UV sequence as compared to the reverse (UV/Cl_2). UV treatment appears to alter the NOM so that it is more reactive with chlorine, producing more TOX, but not measurably more THMs or HAAs.

At UV doses greater than 400 mJ/cm^2, UV treatment caused a noticeable increase of about 10 to 20 percent in chlorine demand. The increase in demand was greater at higher UV doses. DBP concentrations also increased substantially with increasing UV dose. This was especially noticeable with TOX formation.

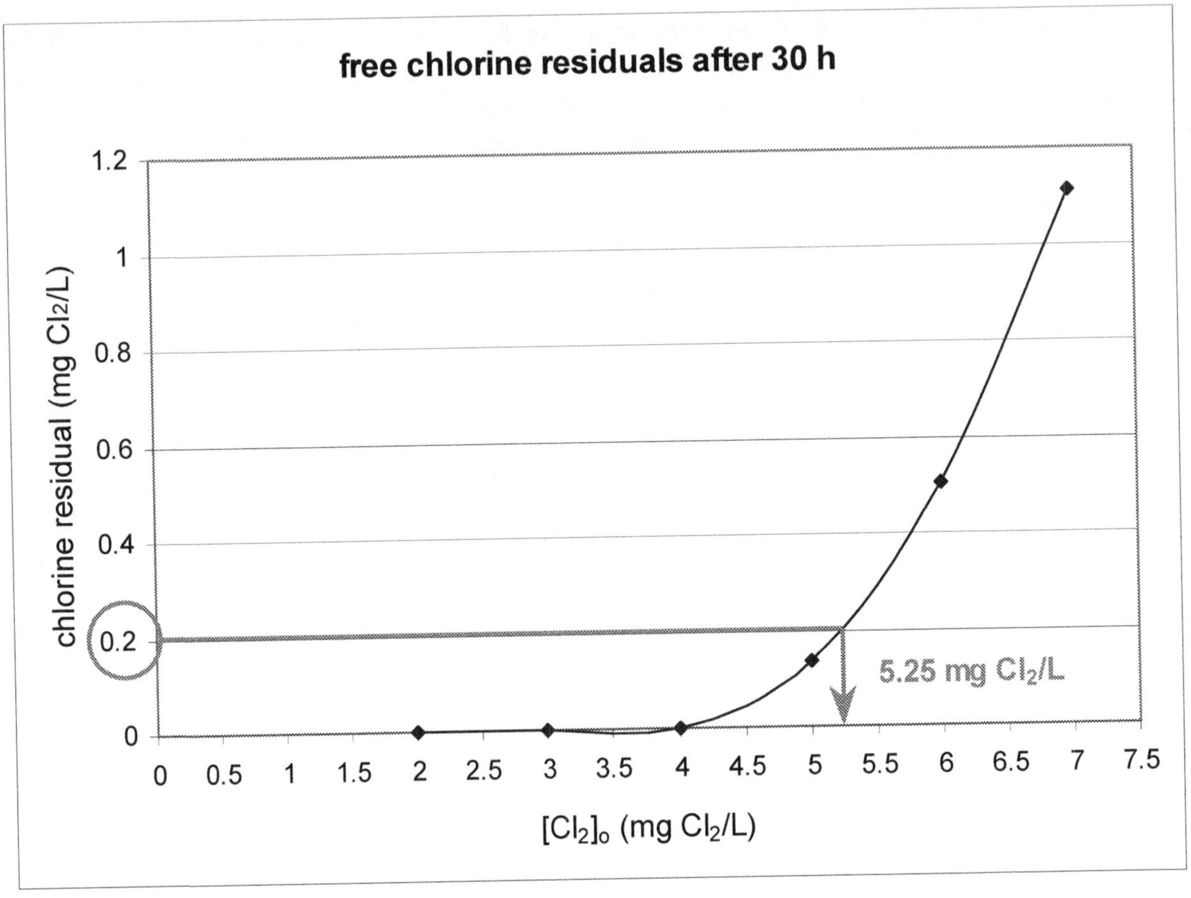

Figure 5.1 Free-chlorine dose test – DBP Test No. 1

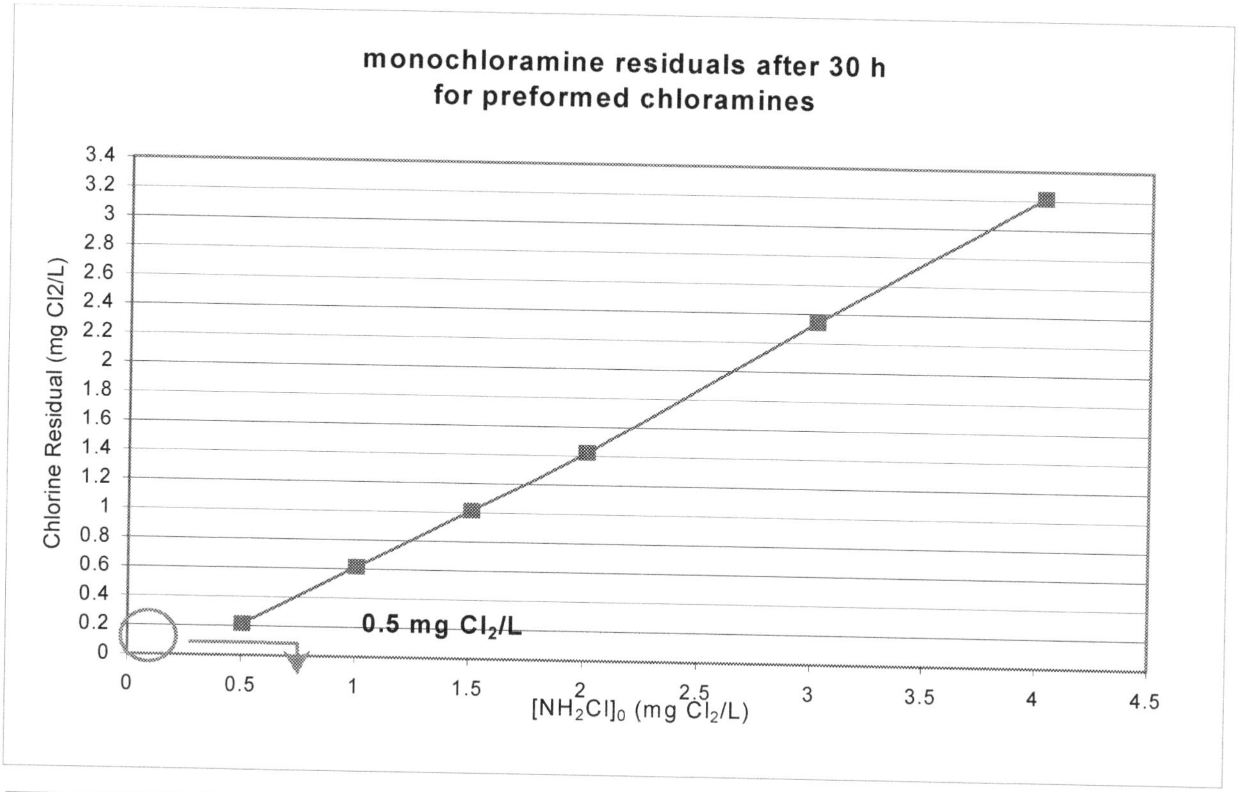

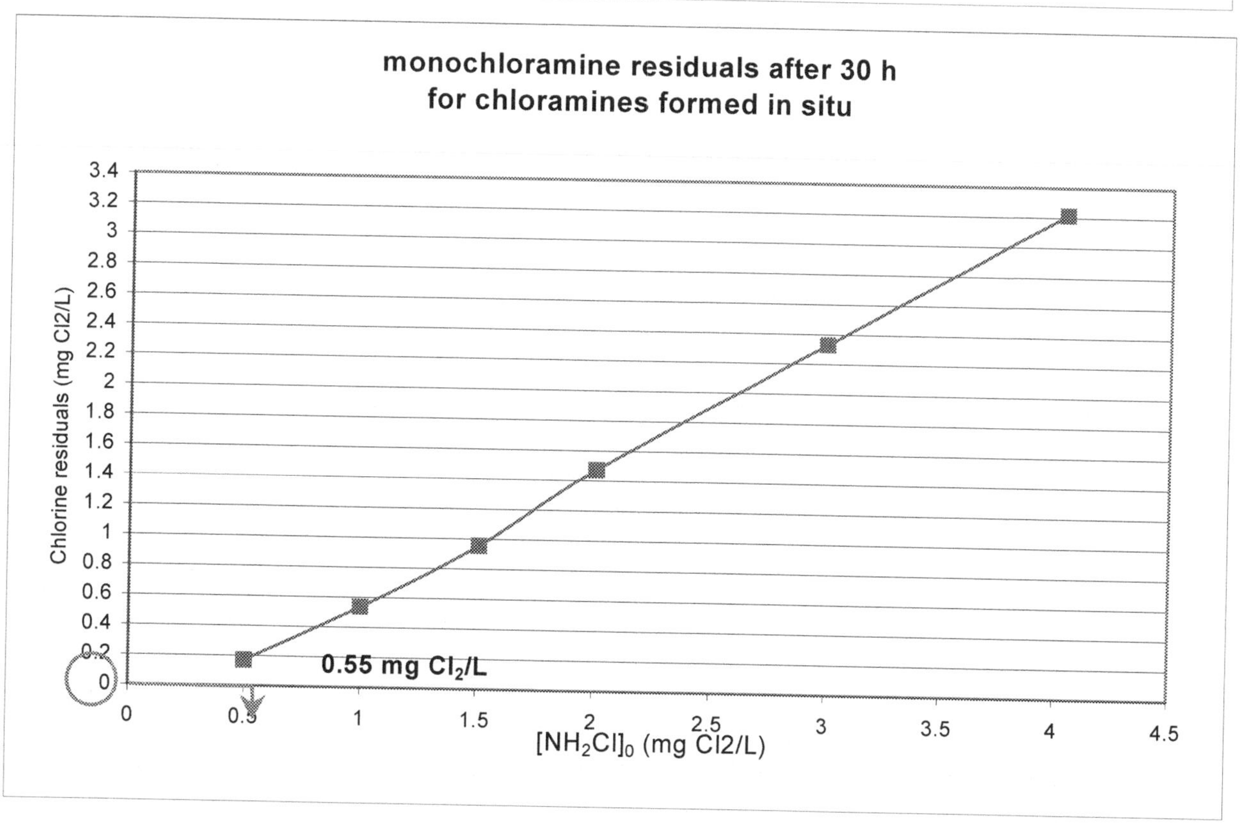

Figure 5.2 Chloramine dose test – DBP Test No. 1 (top – preformed, bottom – in situ)

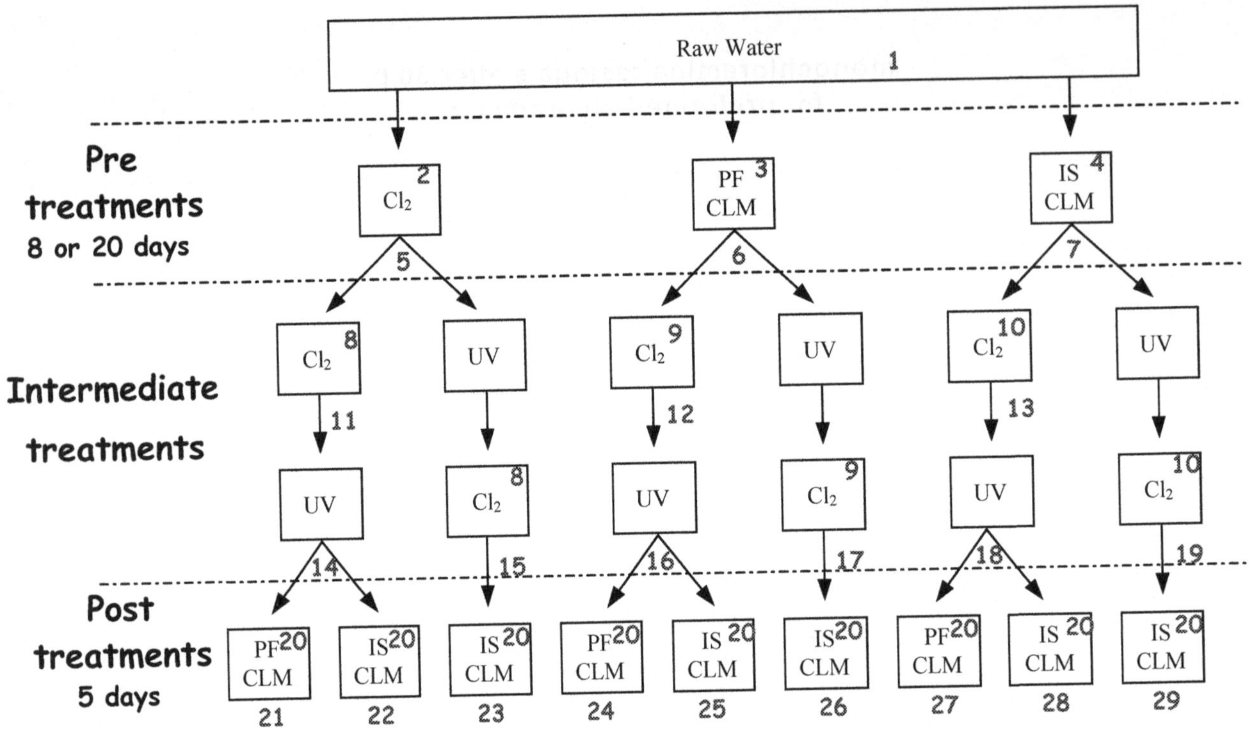

Figure 5.3 Sample number key for chlorine dose/residual data – DBP Test Nos. 1 through 5 (PF-CLM: Preformed chloramines, IS-CLM: In situ formed chloramines)

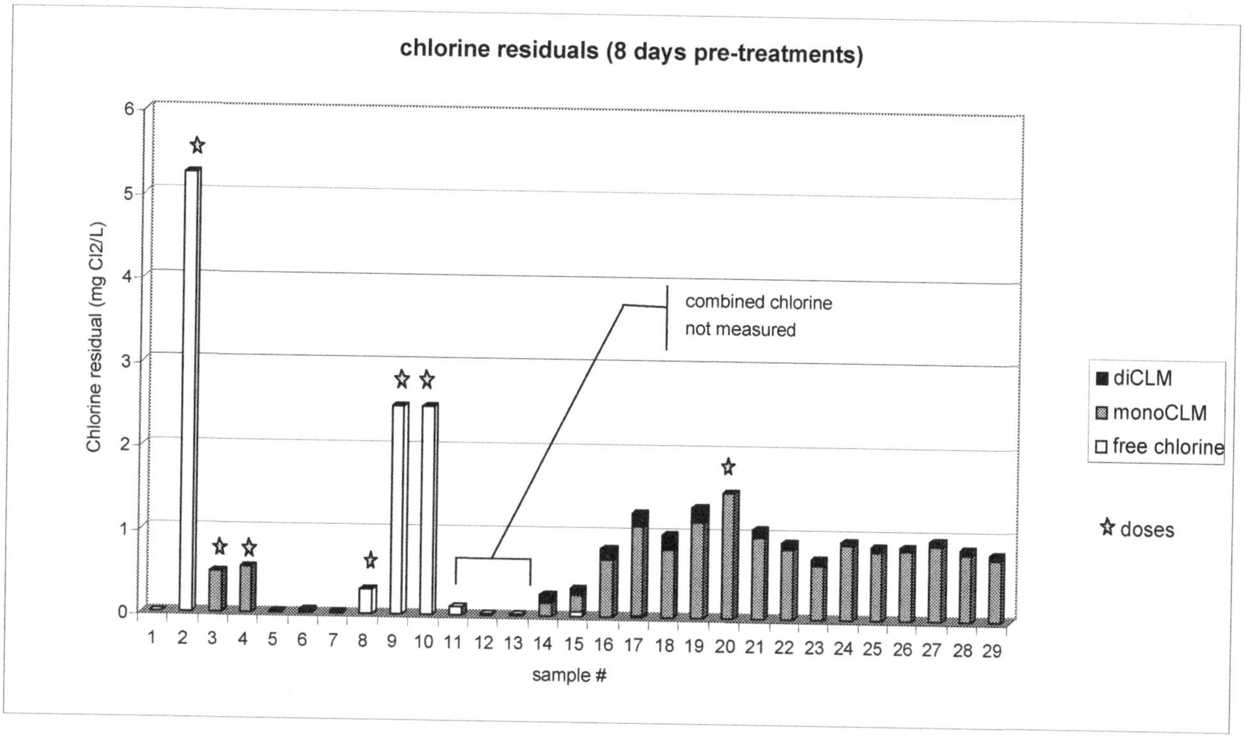

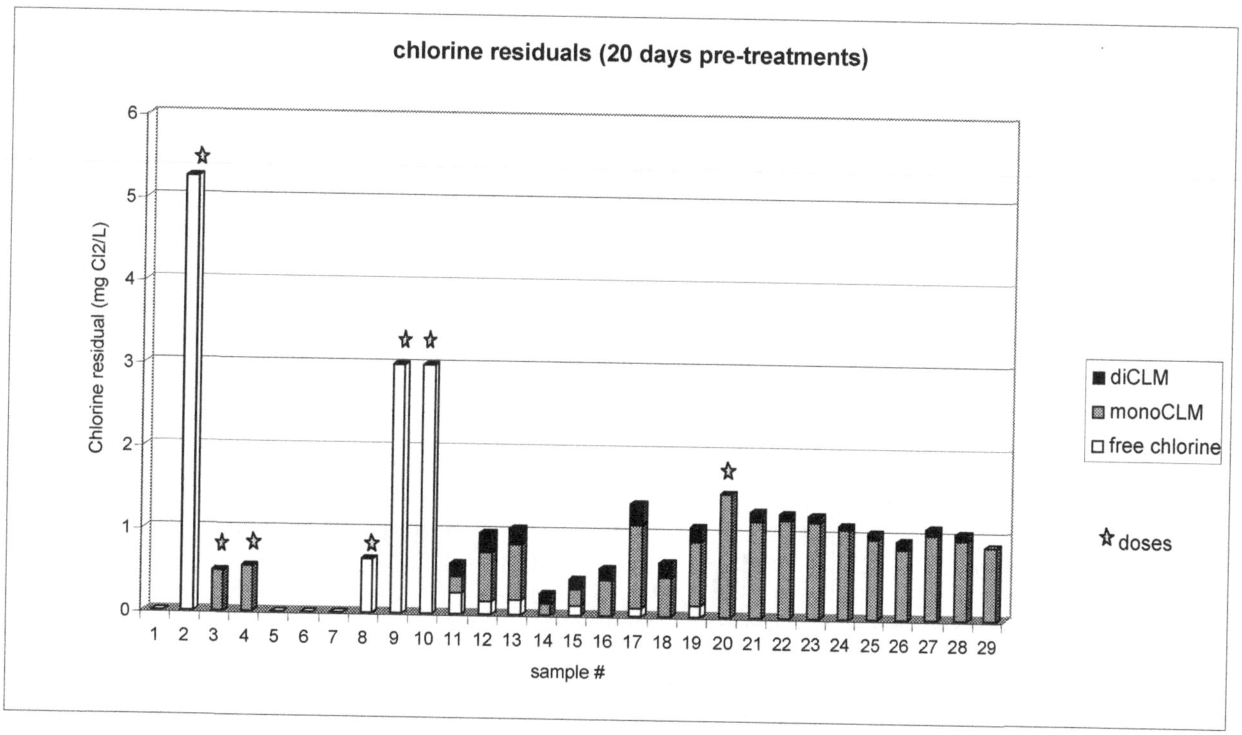

Figure 5.4 Chlorine doses and residuals for 8-day and 20-day pretreatments DBP Test No. 1

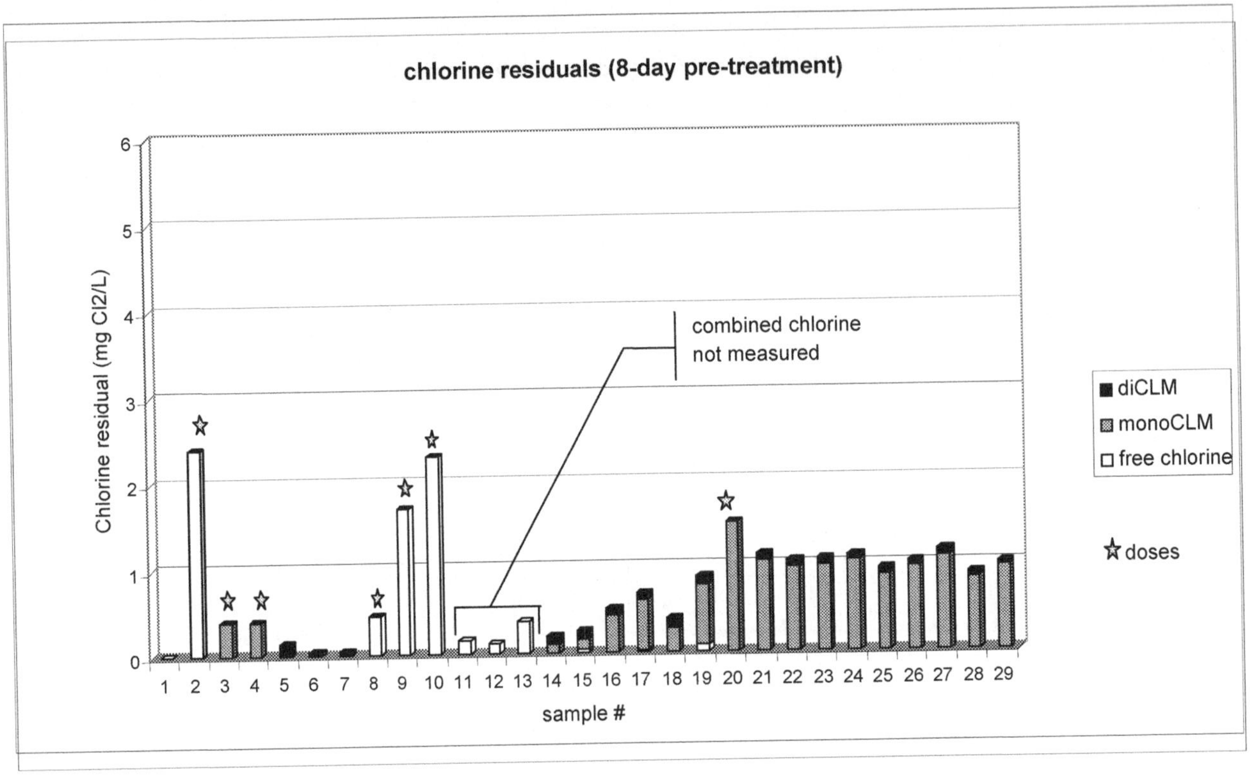

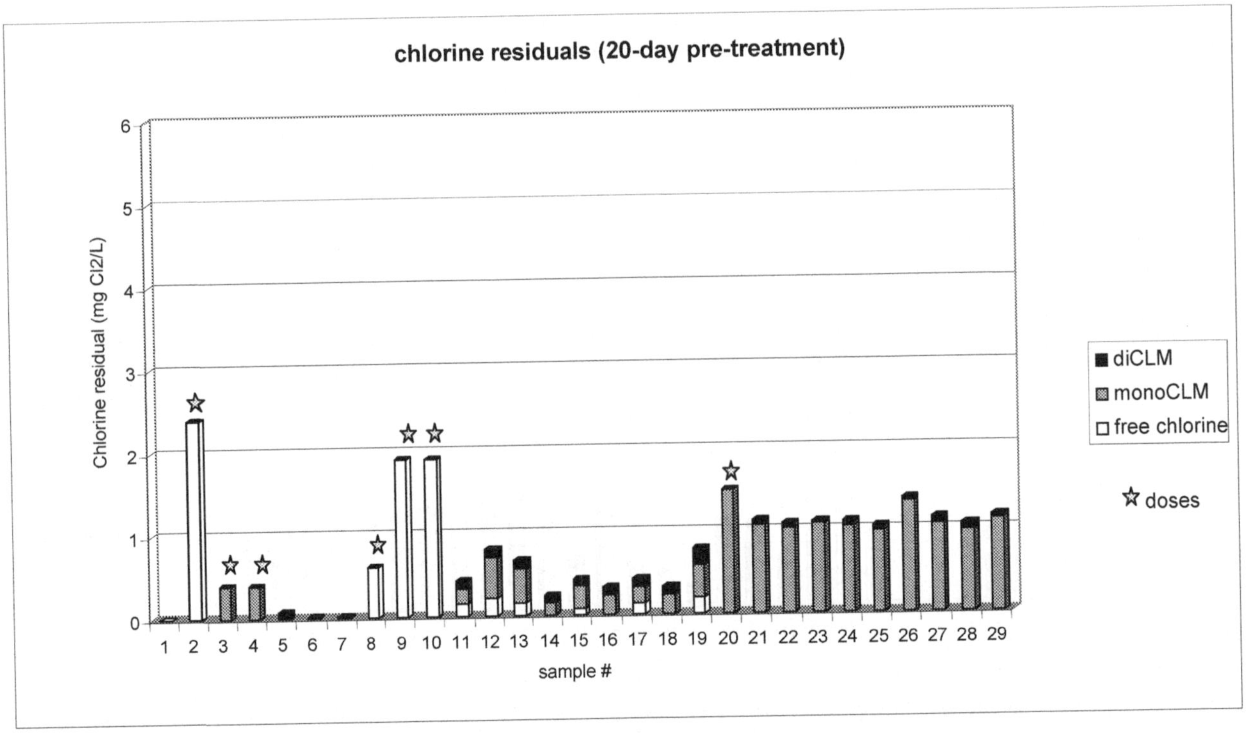

Figure 5.5 Chlorine doses and residuals for 8-day and 20-day pretreatments DBP Test No. 2

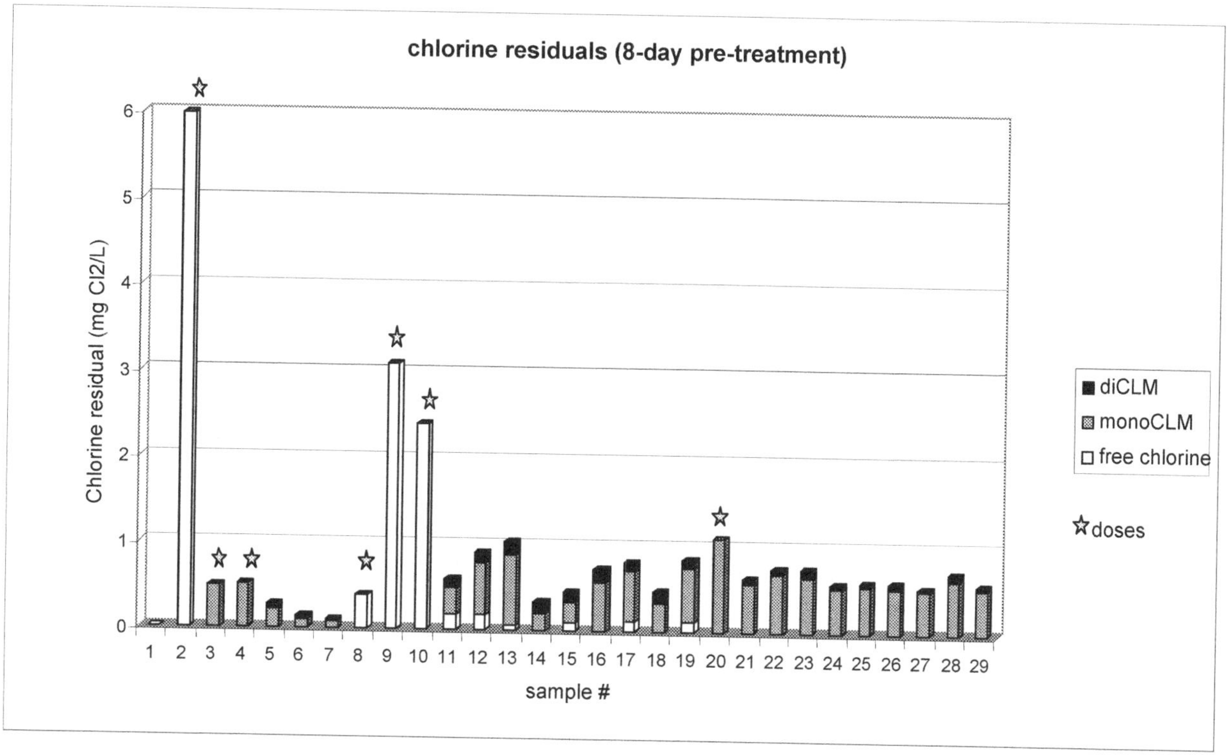

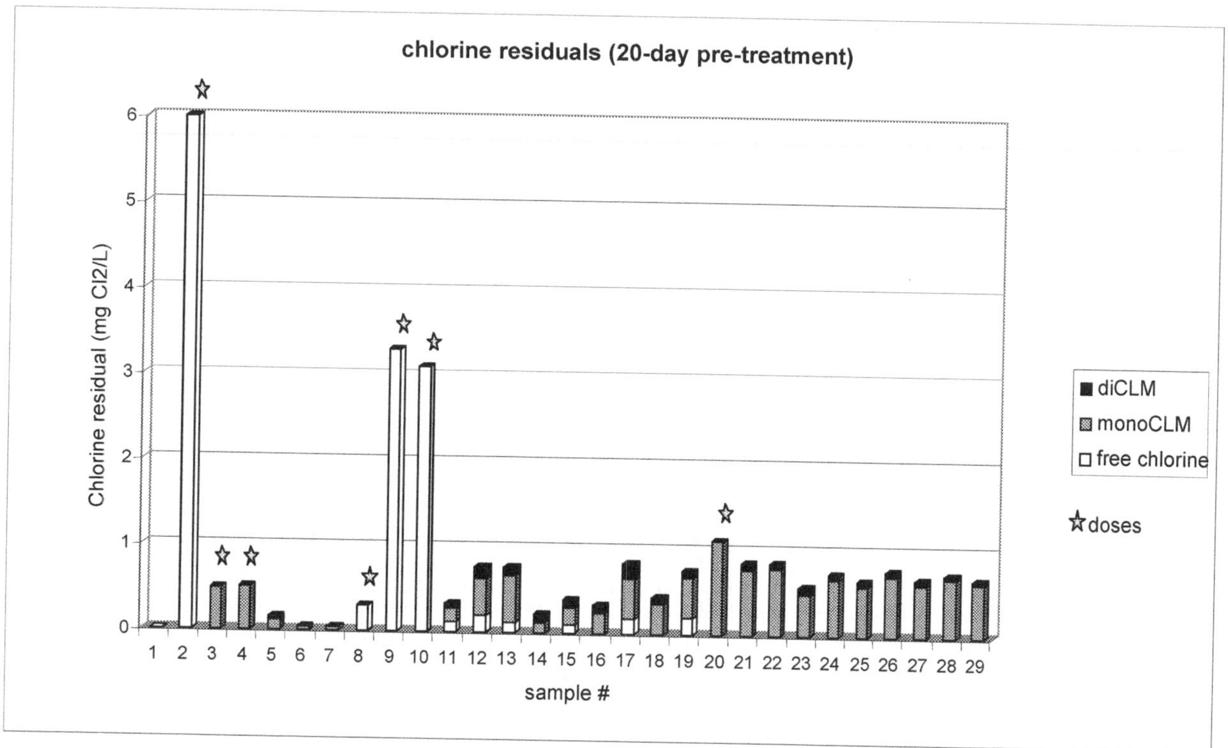

Figure 5.6 Chlorine doses and residuals for 8-day and 20-day pretreatments DBP Test No. 3

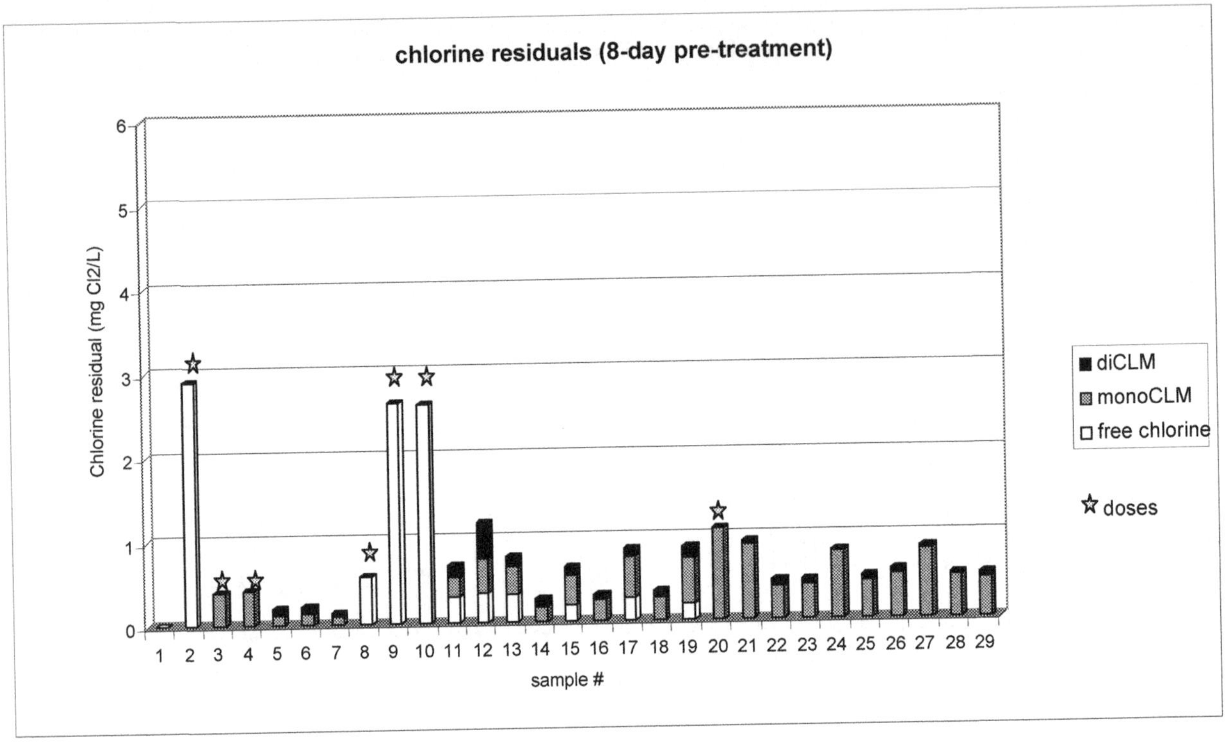

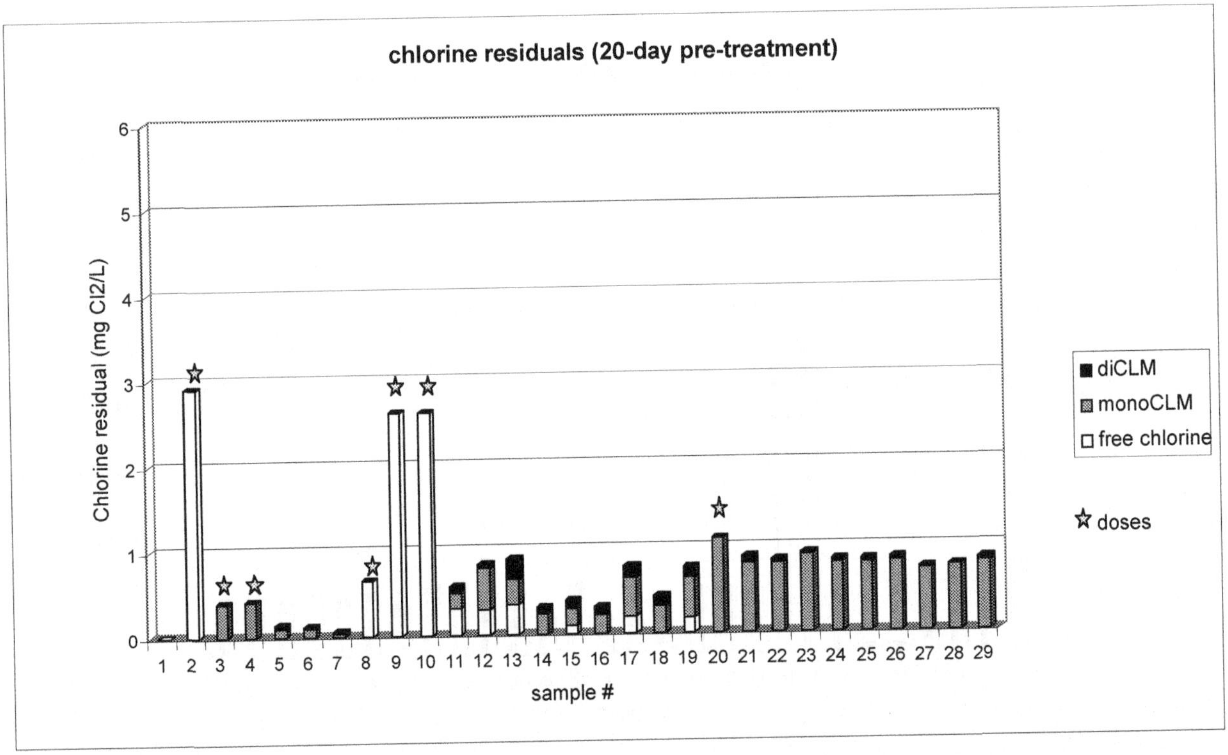

Figure 5.7 Chlorine doses and residuals for 8-day and 20-day pretreatments DBP Test No. 4

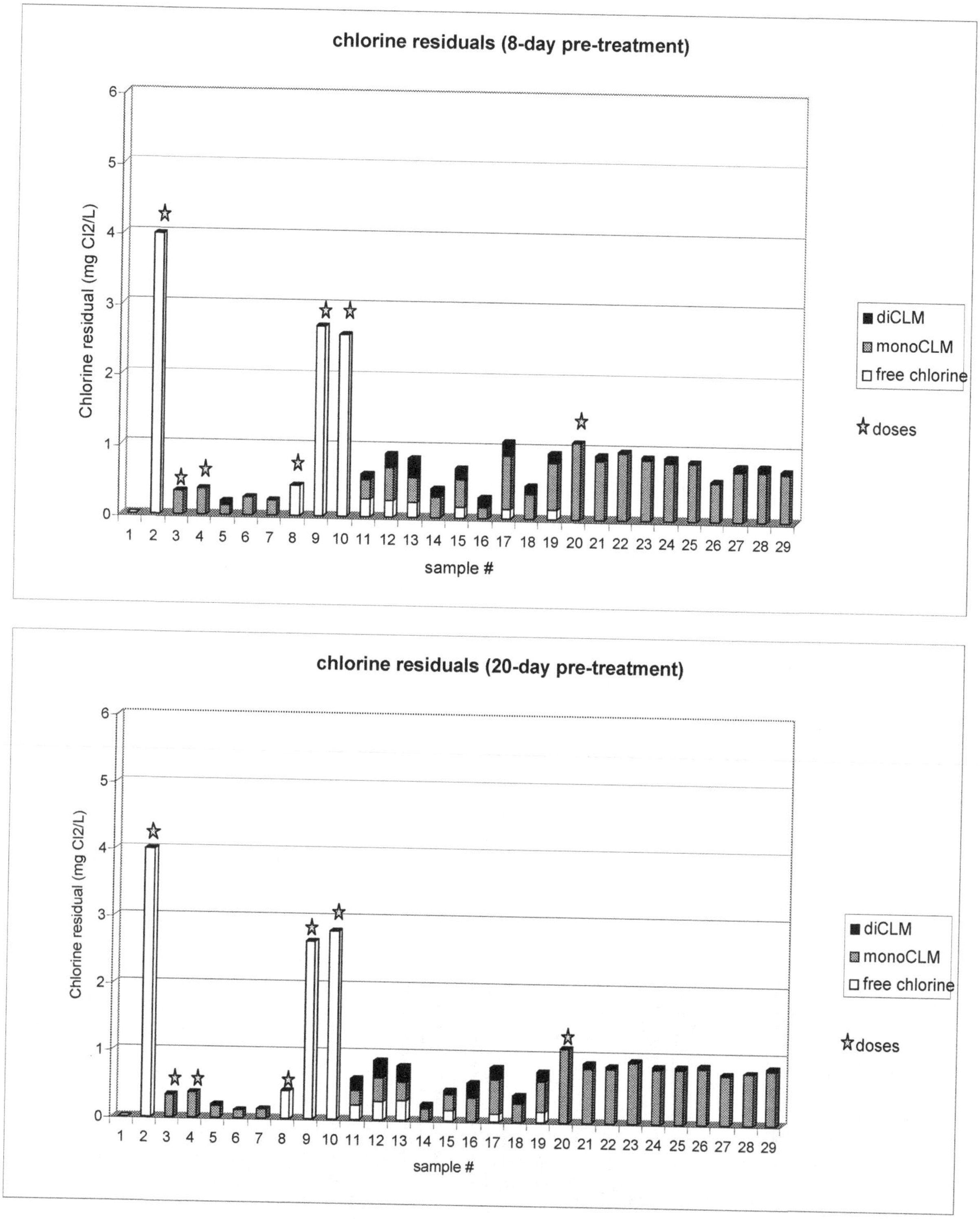

Figure 5.8 Chlorine doses and residuals for 8-day and 20-day pretreatments DBP Test No. 5

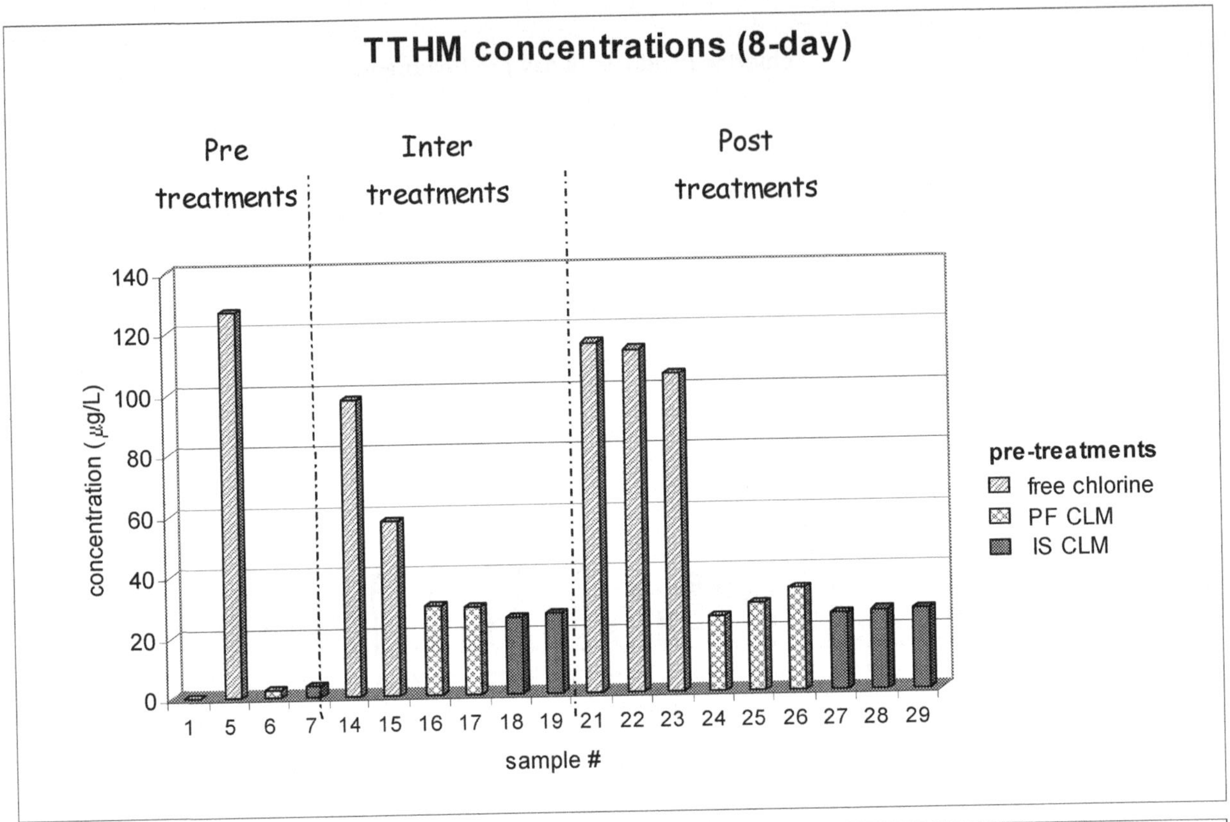

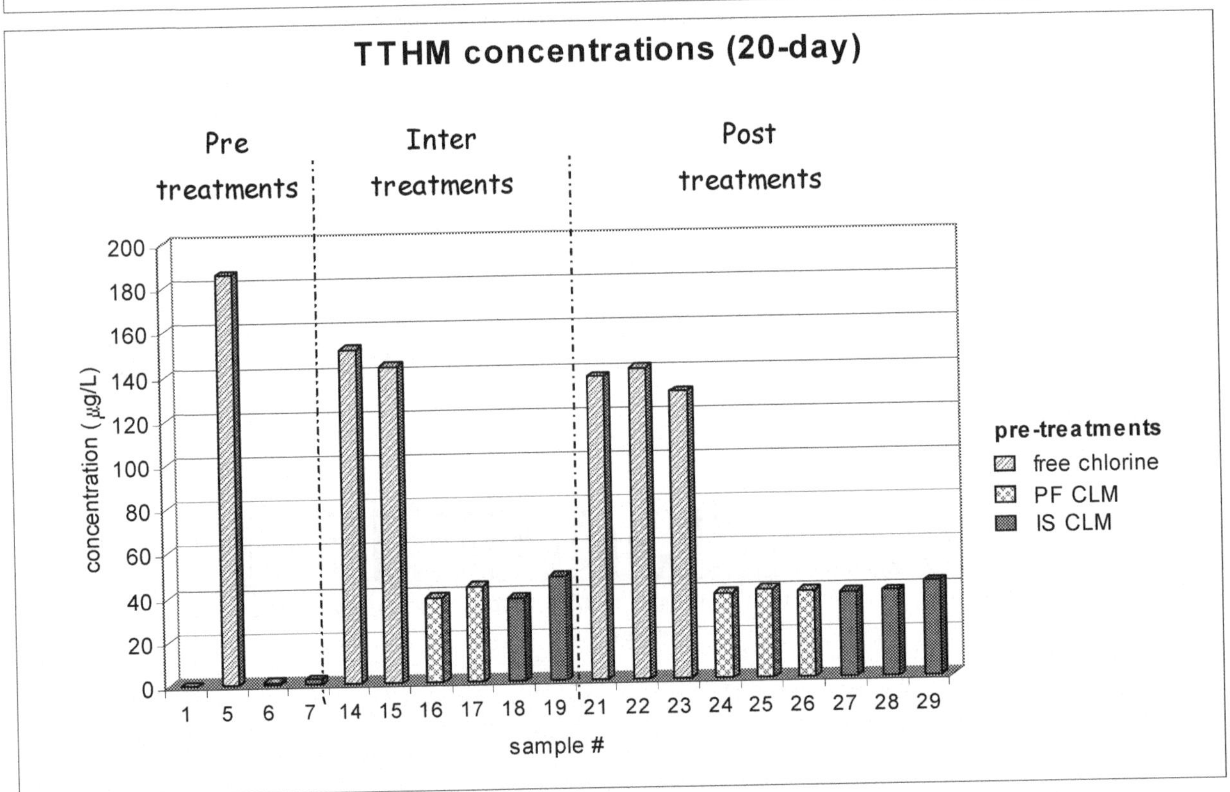

Figure 5.9 TTHM results for 8-day and 20-day pretreatment tests – DBP Test No. 1 – June 8, 2001

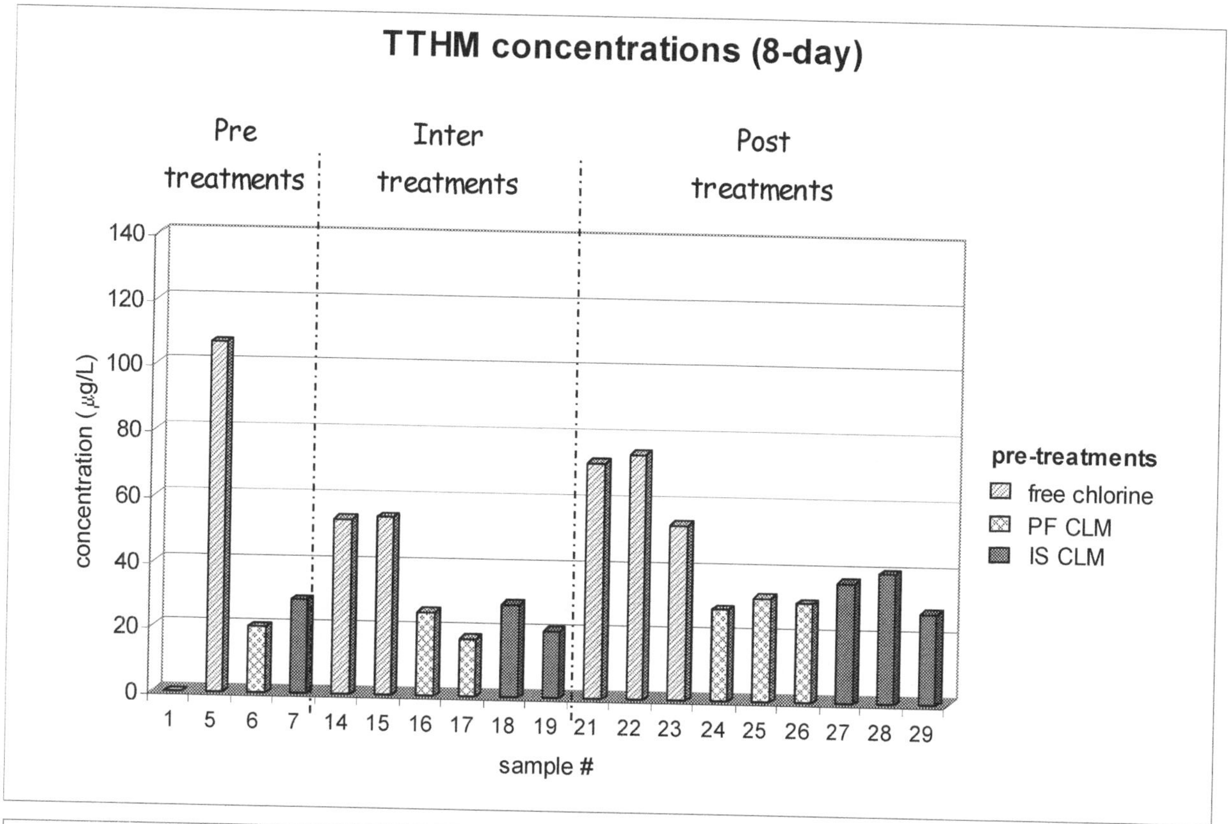

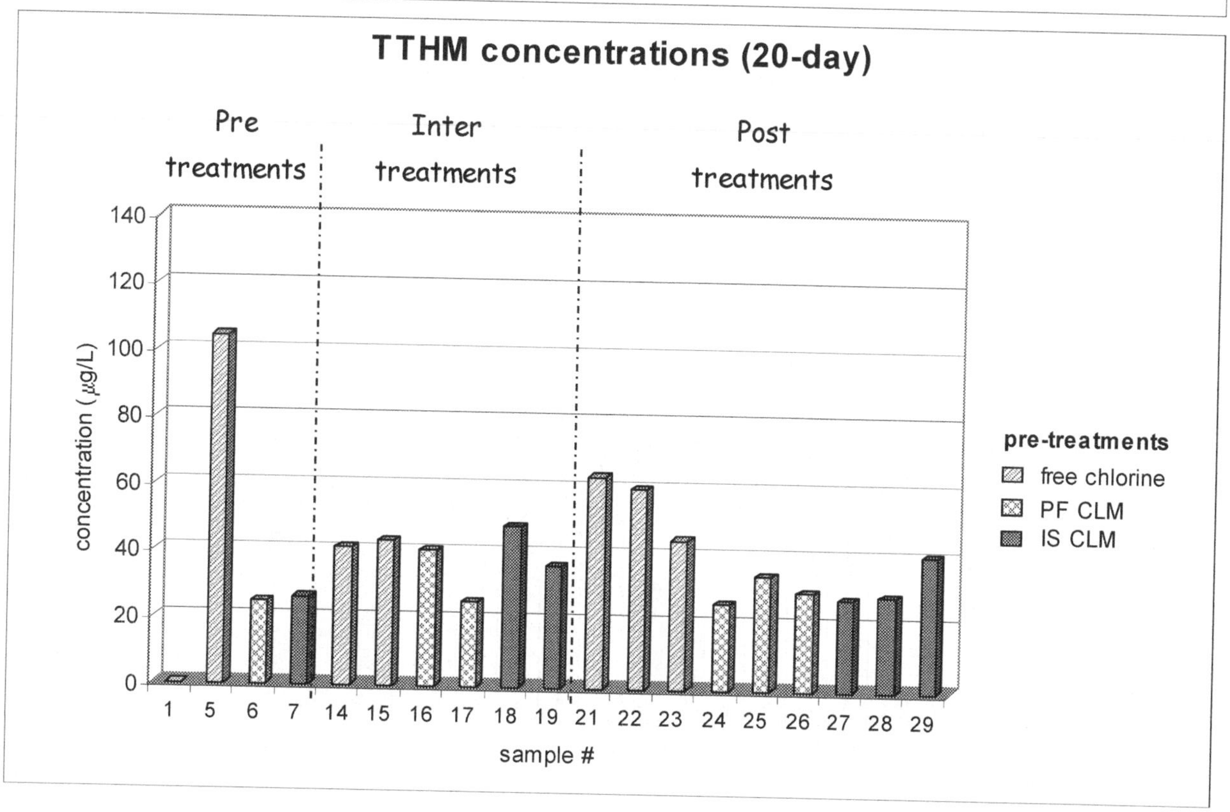

Figure 5.10 TTHM results for 8-day and 20-day pretreatment tests – DBP Test No. 2 – July 16, 2001

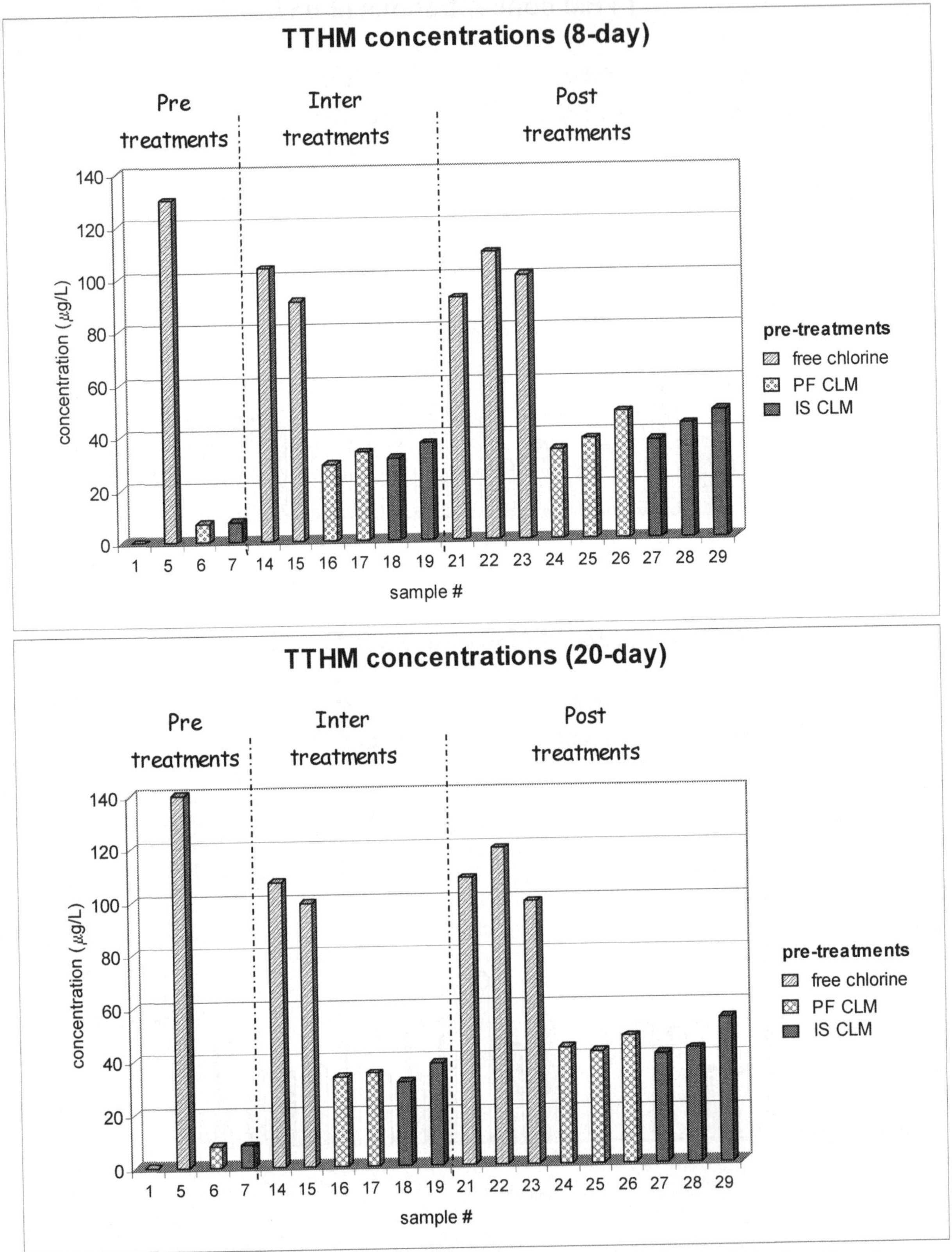

Figure 5.11 TTHM results for 8-day and 20-day pretreatment tests – DBP Test No. 3 – October 10, 2001

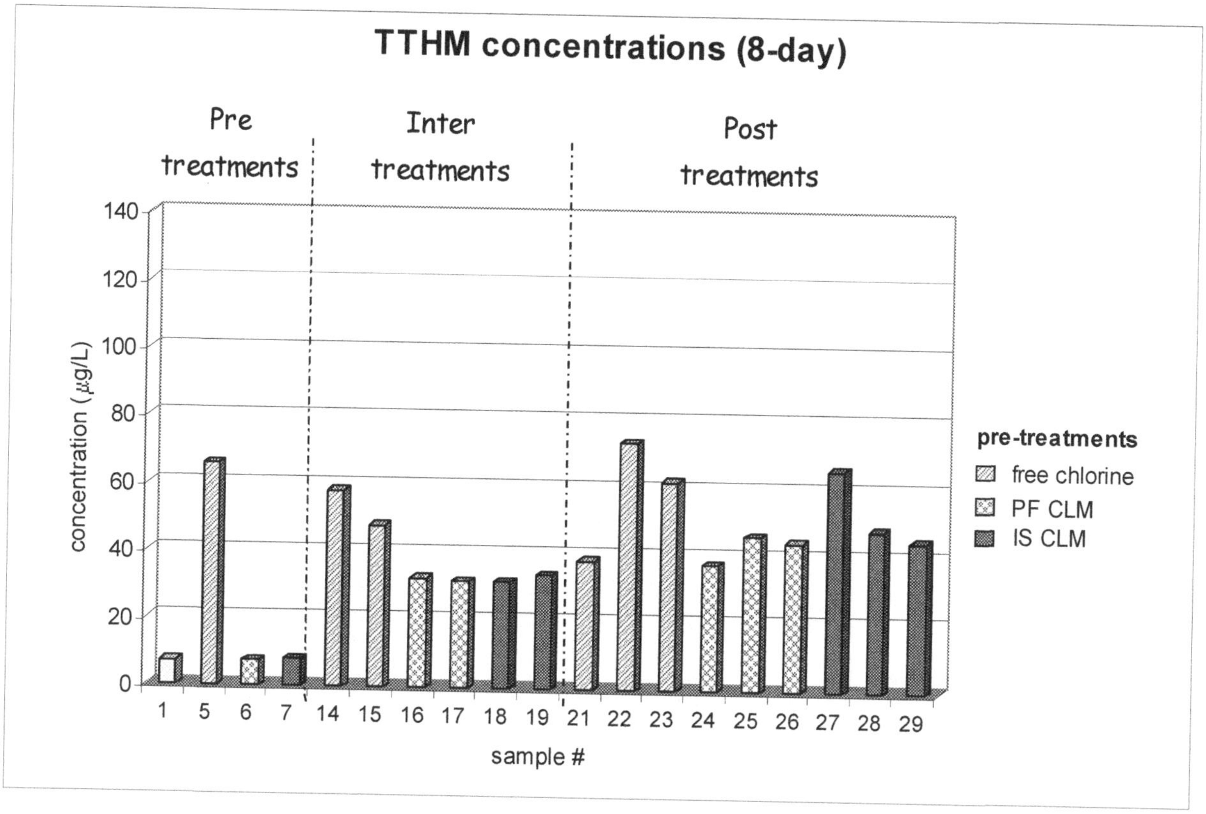

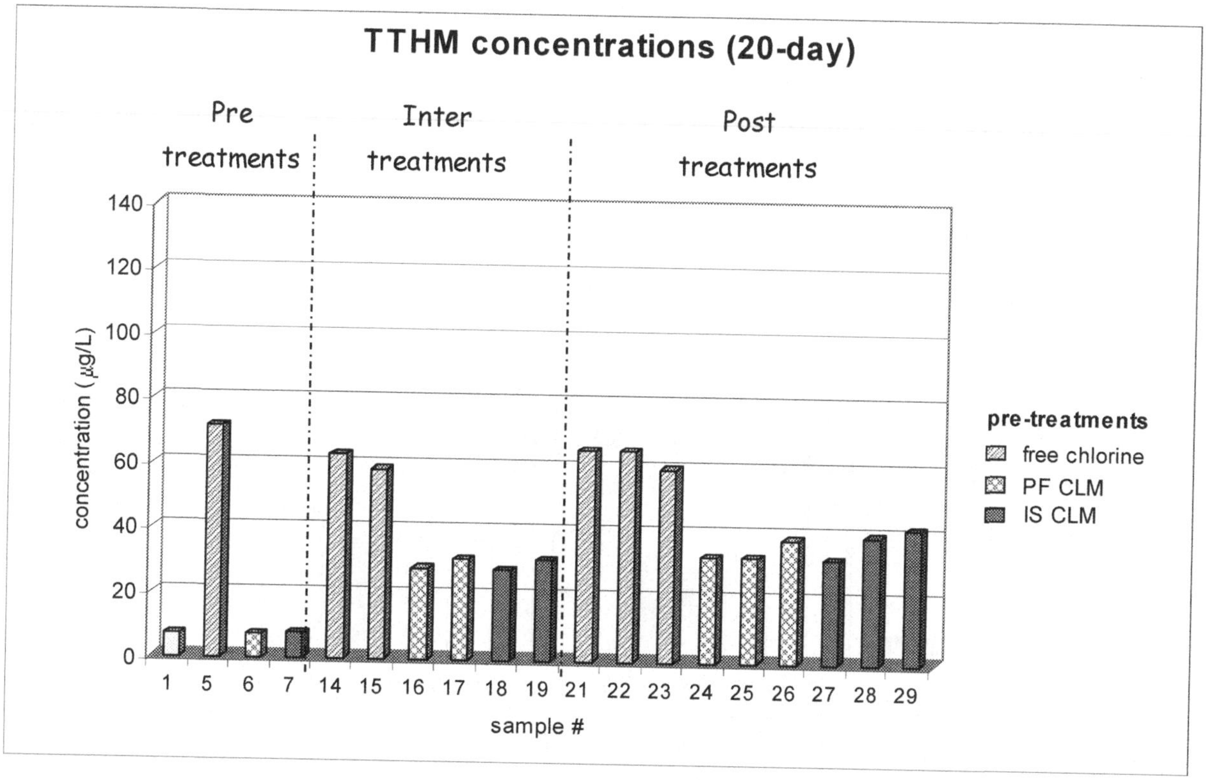

Figure 5.12 TTHM results for 8-day and 20-day pretreatment tests – DBP Test No. 4 – November 26, 2001

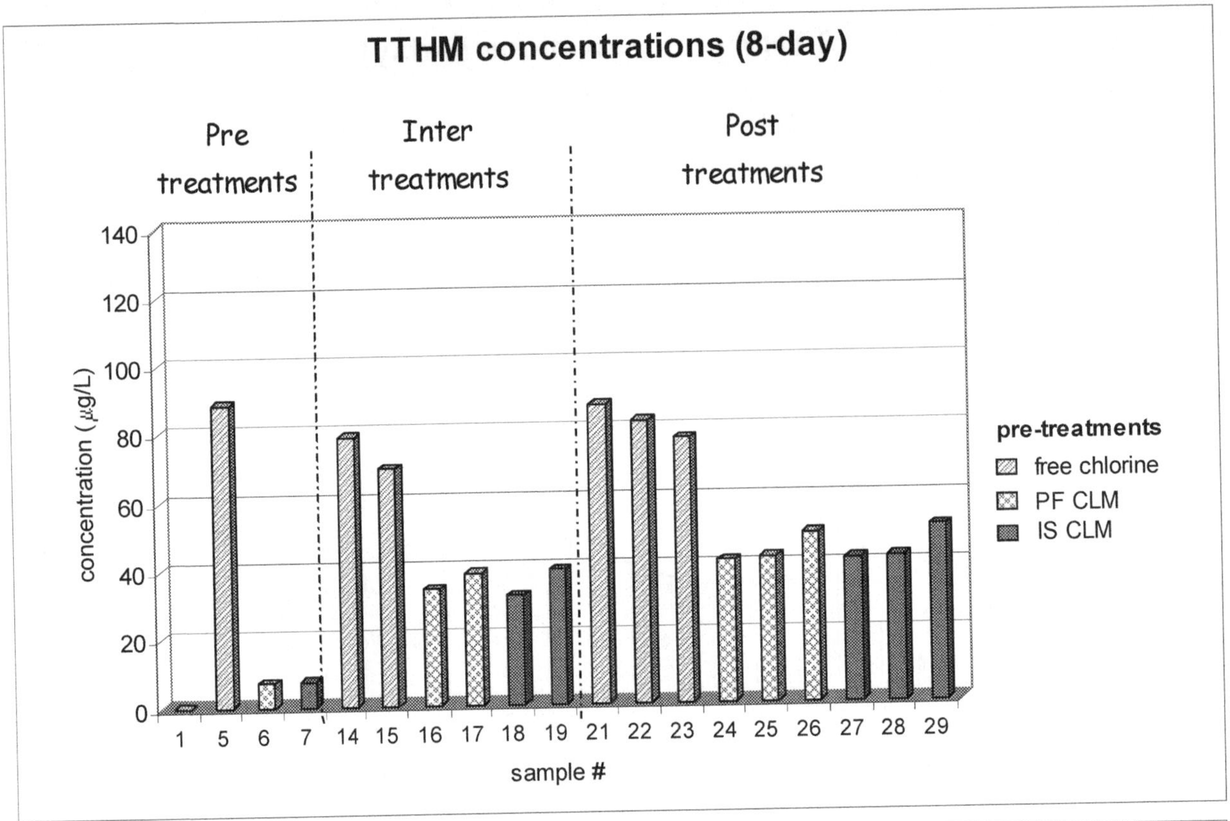

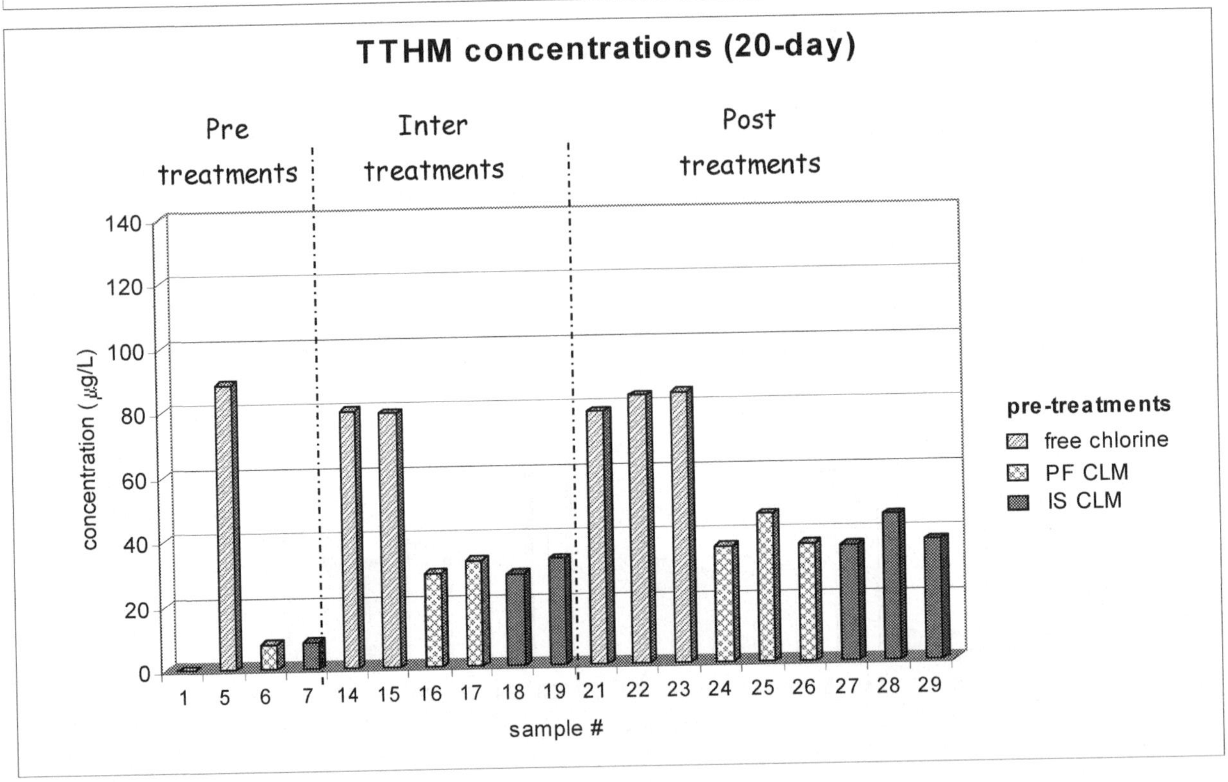

Figure 5.13 TTHM results for 8-day and 20-day pretreatment tests – DBP Test No. 5 – January 30, 2002

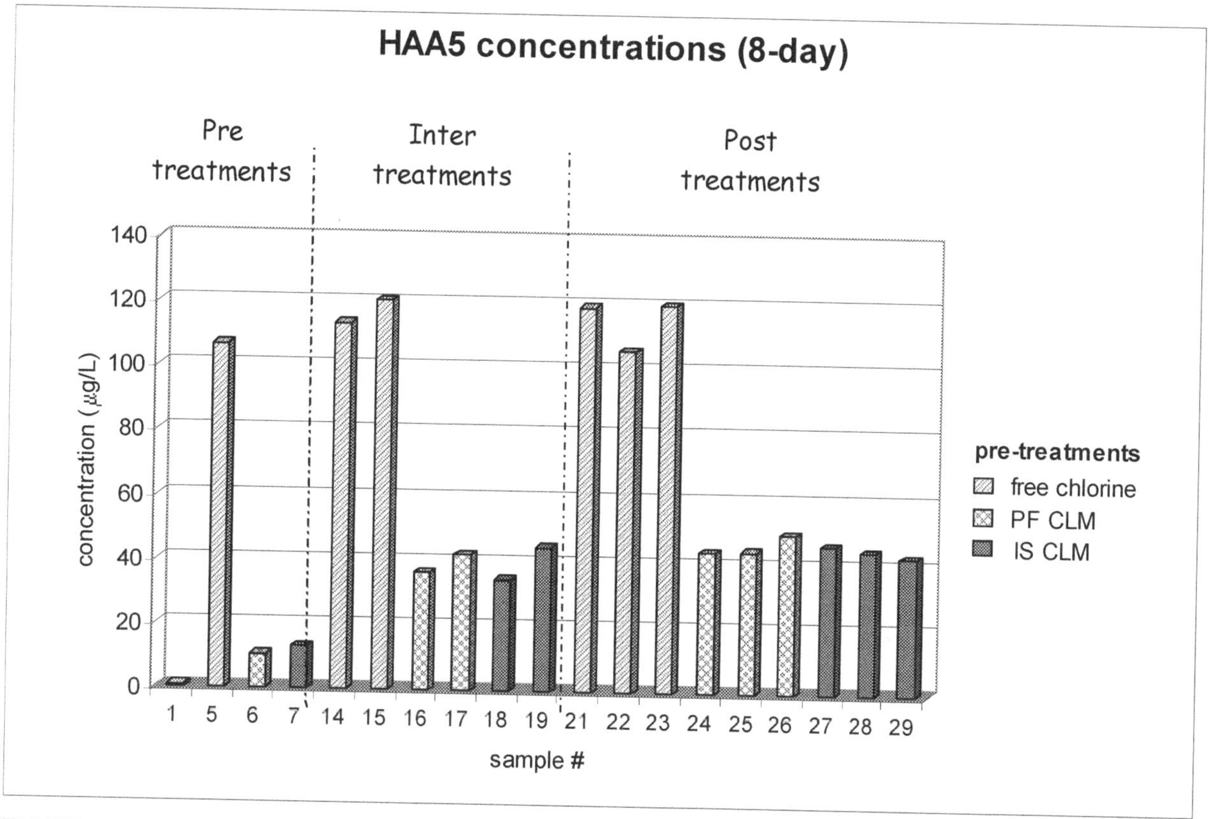

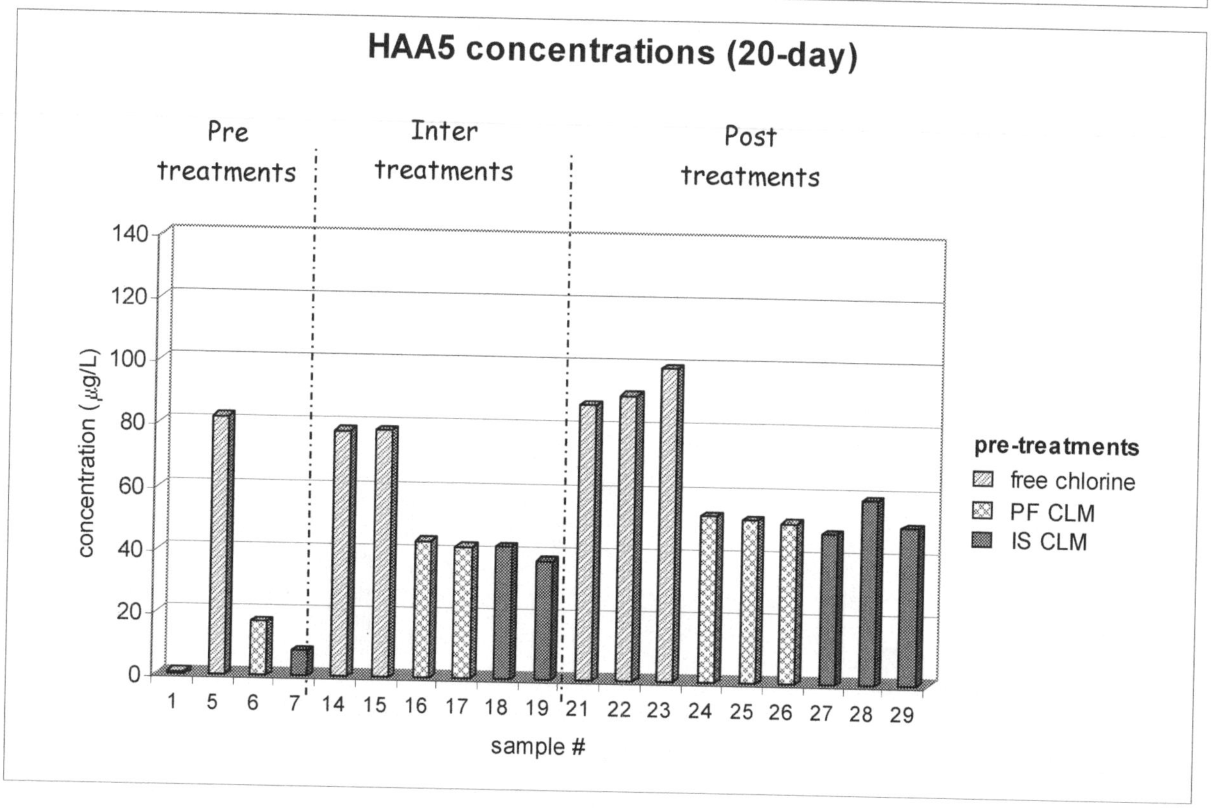

Figure 5.14 HAA results for 8-day and 20-day pretreatment tests – DBP Test No. 1 – June 8, 2001

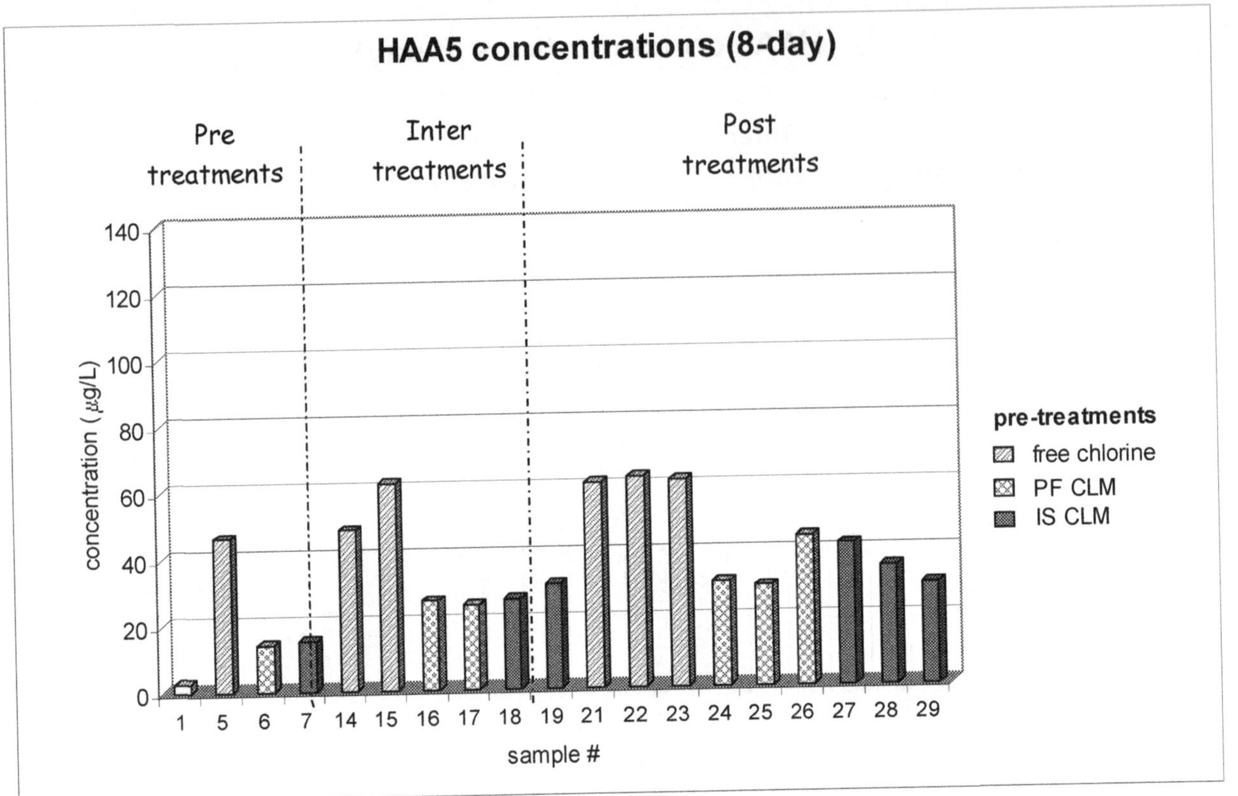

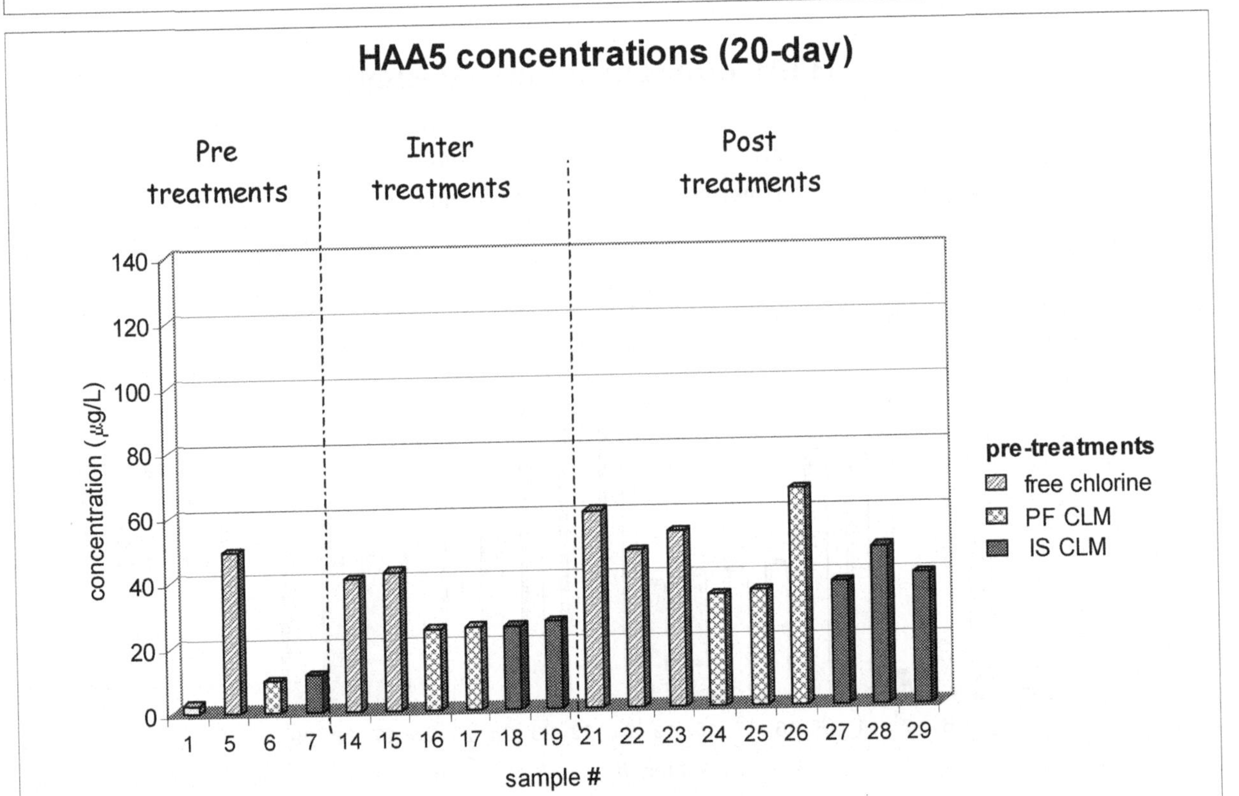

Figure 5.15 HAA results for 8-day and 20-day pretreatment tests – DBP Test No. 2 – July 16, 2001

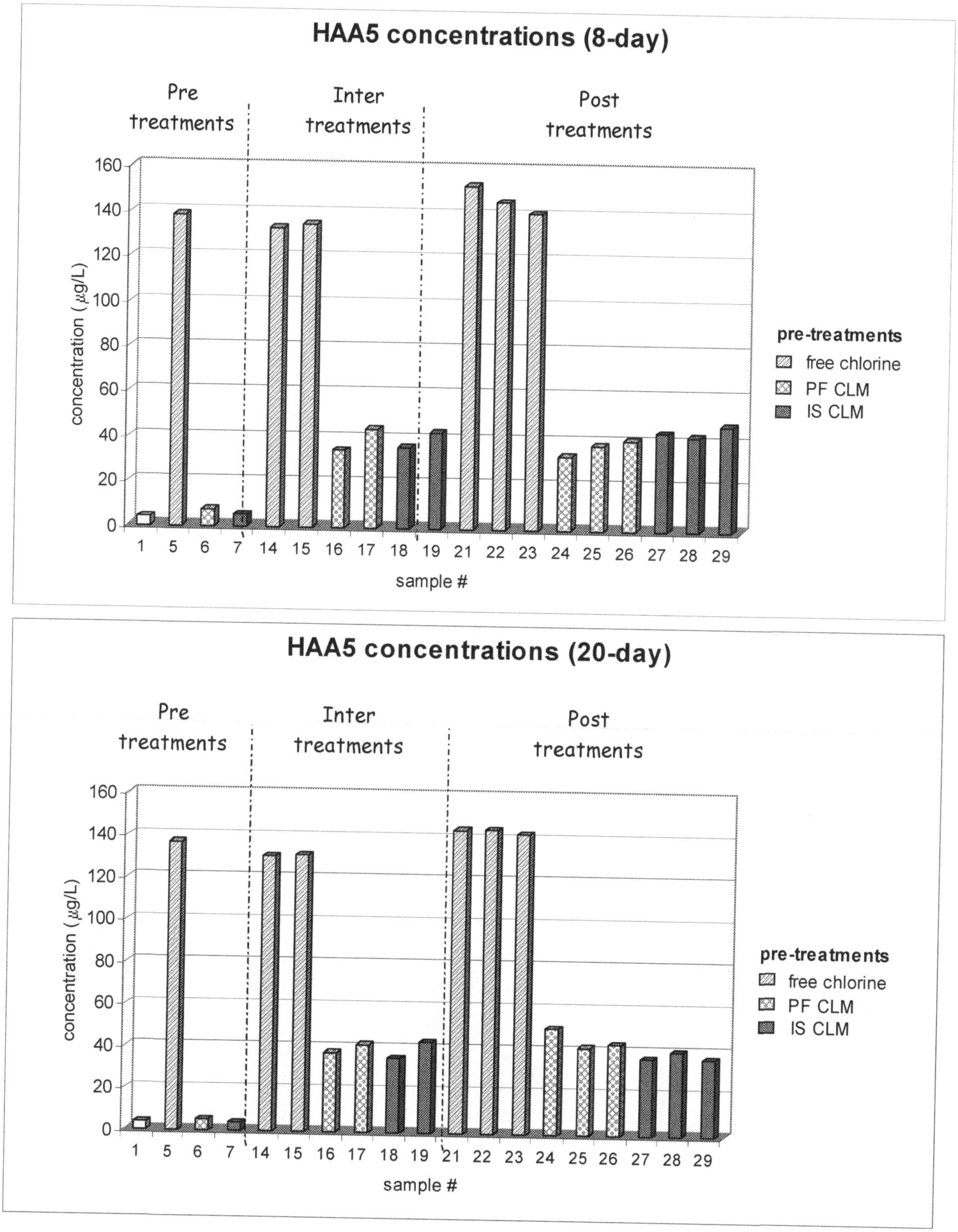

Figure 5.16 HAA results for 8-day and 20-day pretreatment tests – DBP Test No. 3 – October 10, 2001

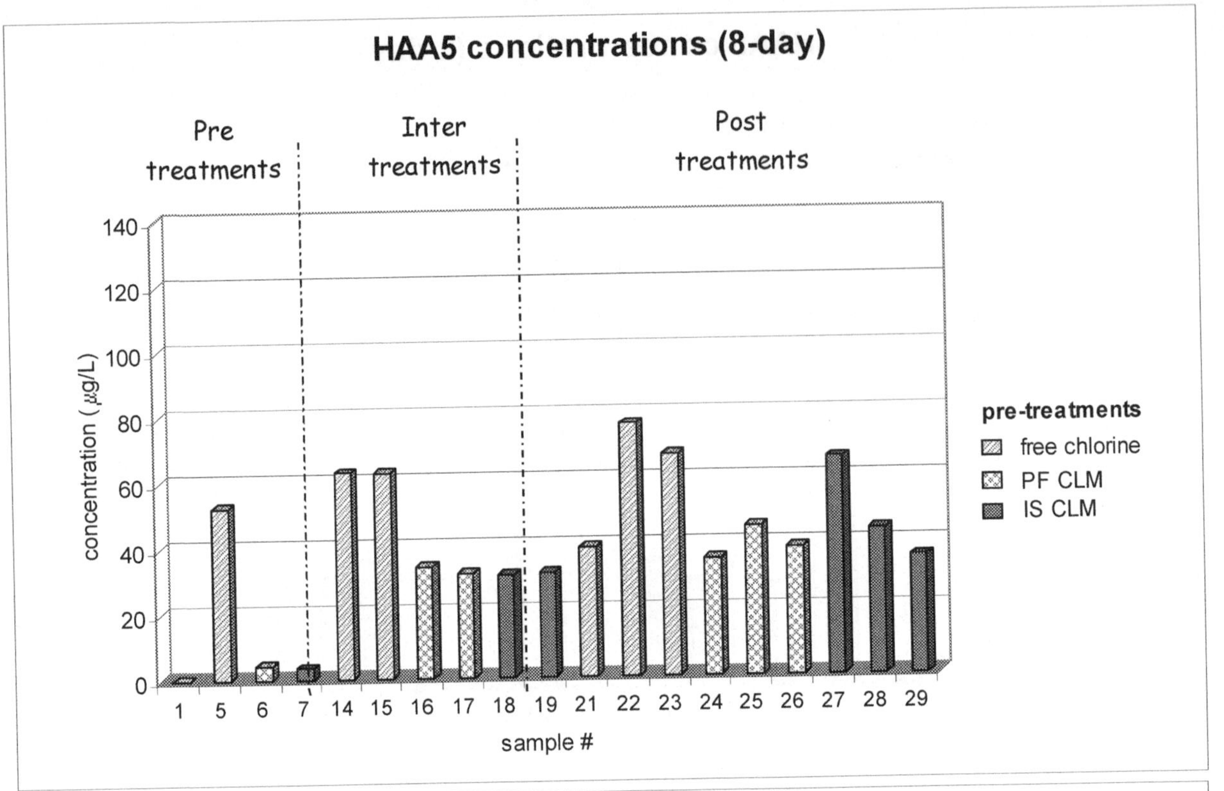

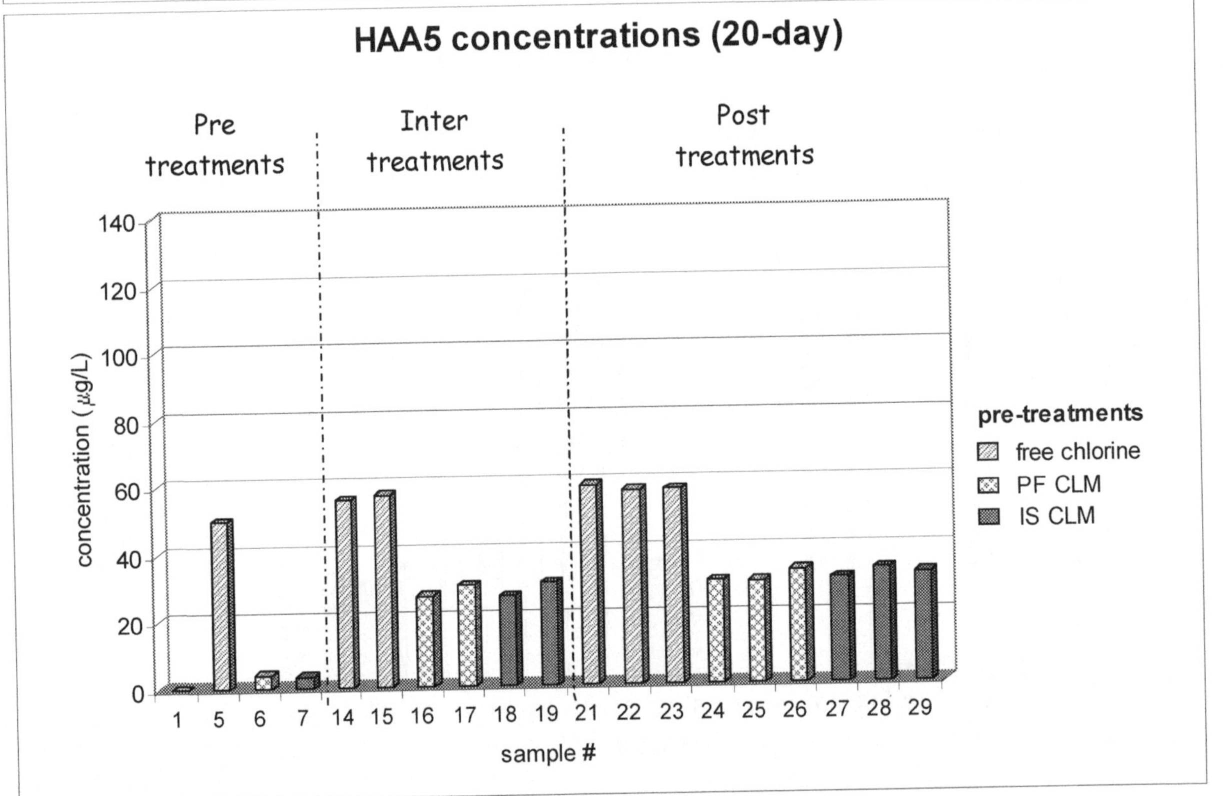

Figure 5.17 HAA results for 8-day and 20-day pretreatment tests – DBP Test No. 4 – November 26, 2001

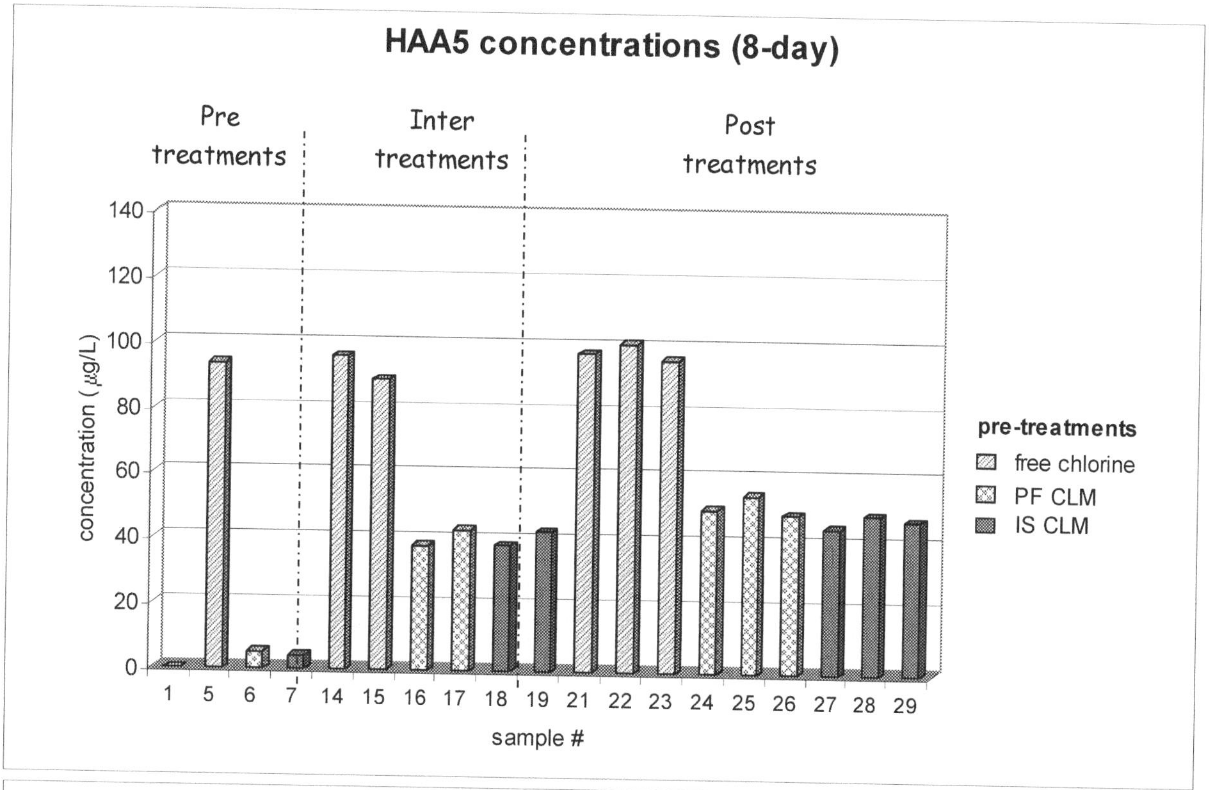

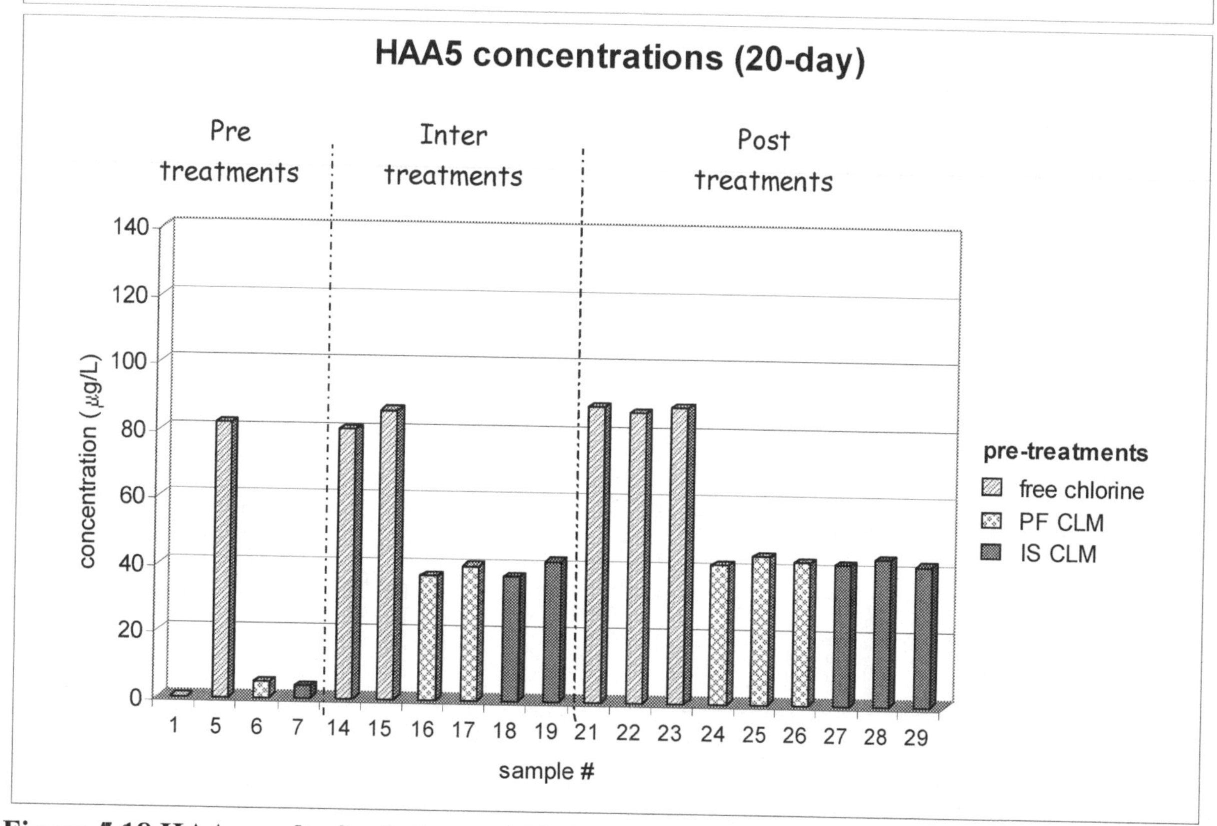

Figure 5.18 HAA results for 8-day and 20-day pretreatment tests – DBP Test No. 5 – January 30, 2002

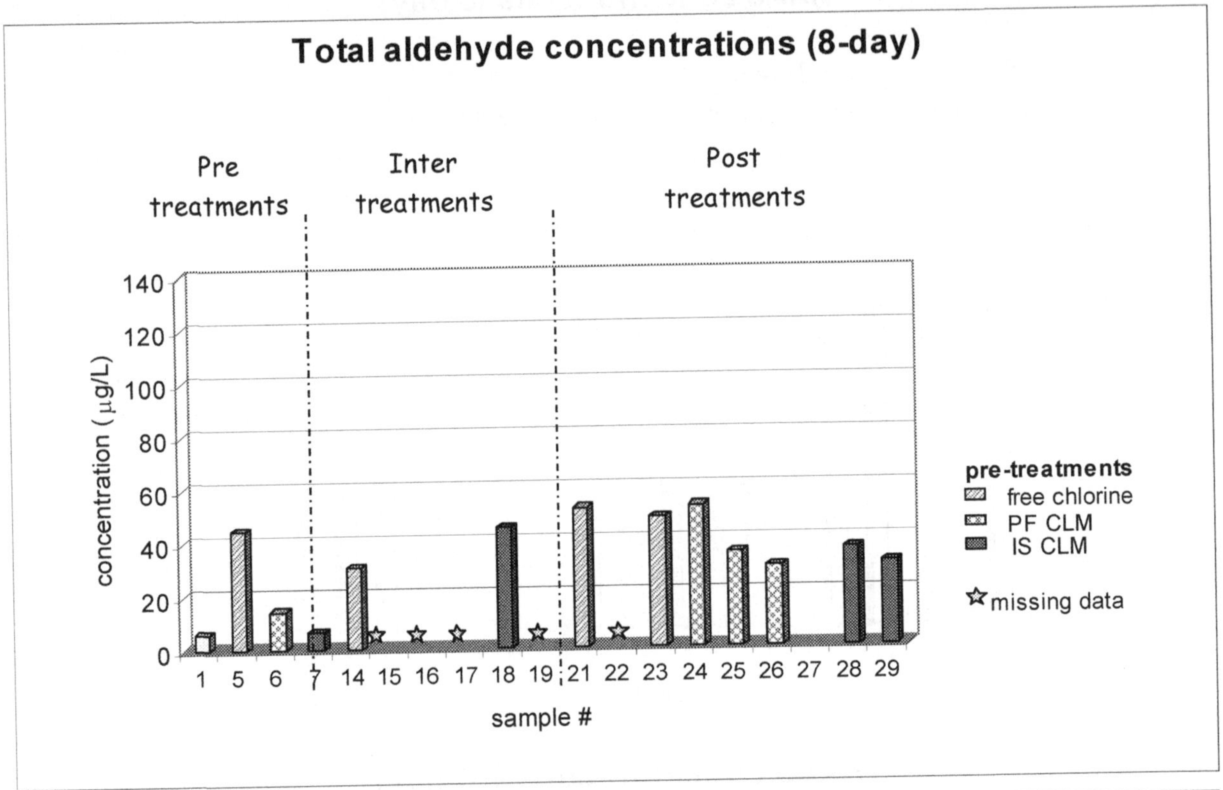

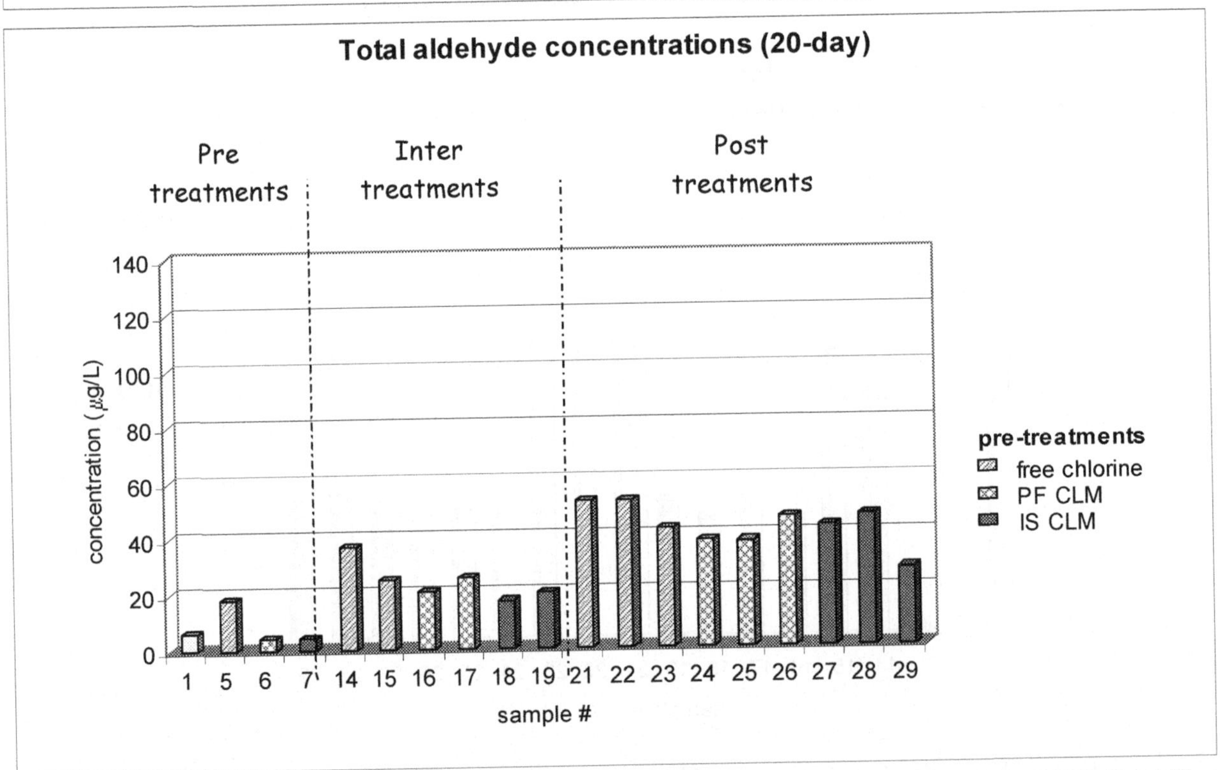

Figure 5.19 Total aldehyde results for 8-day and 20-day pretreatment tests – DBP Test No. 1 – June 8, 2001

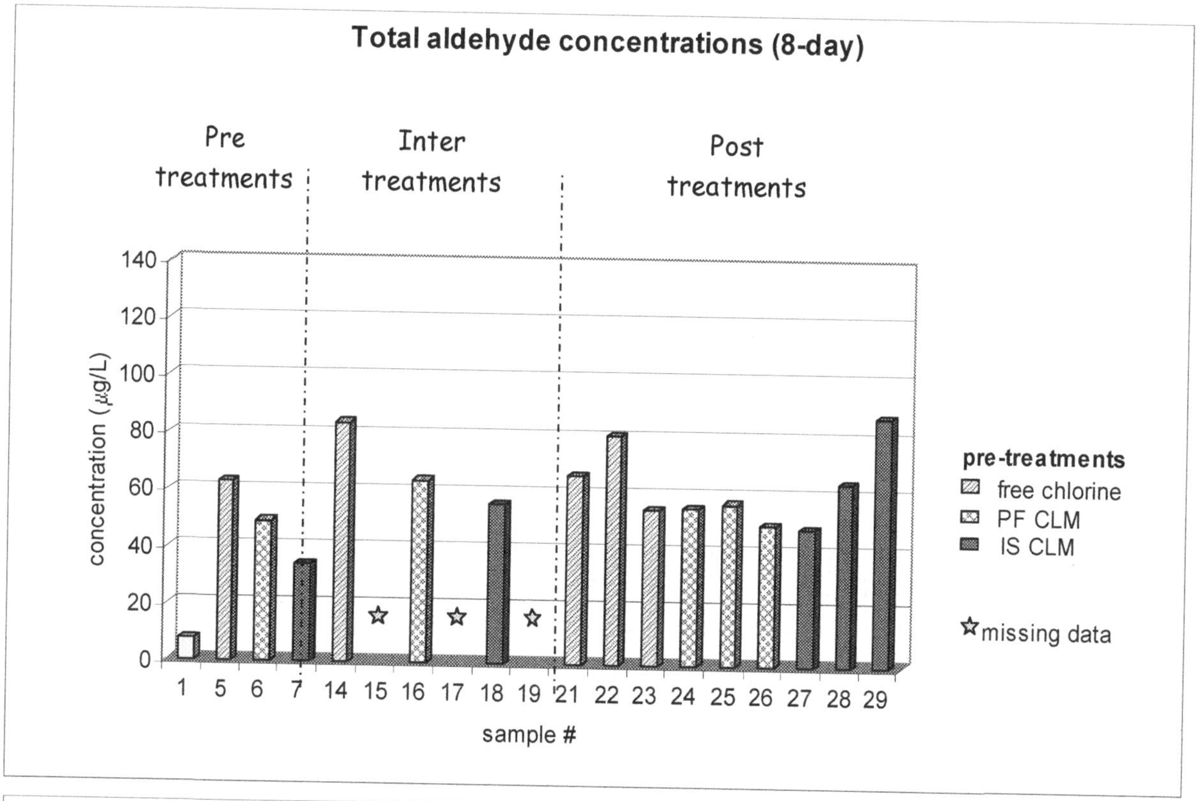

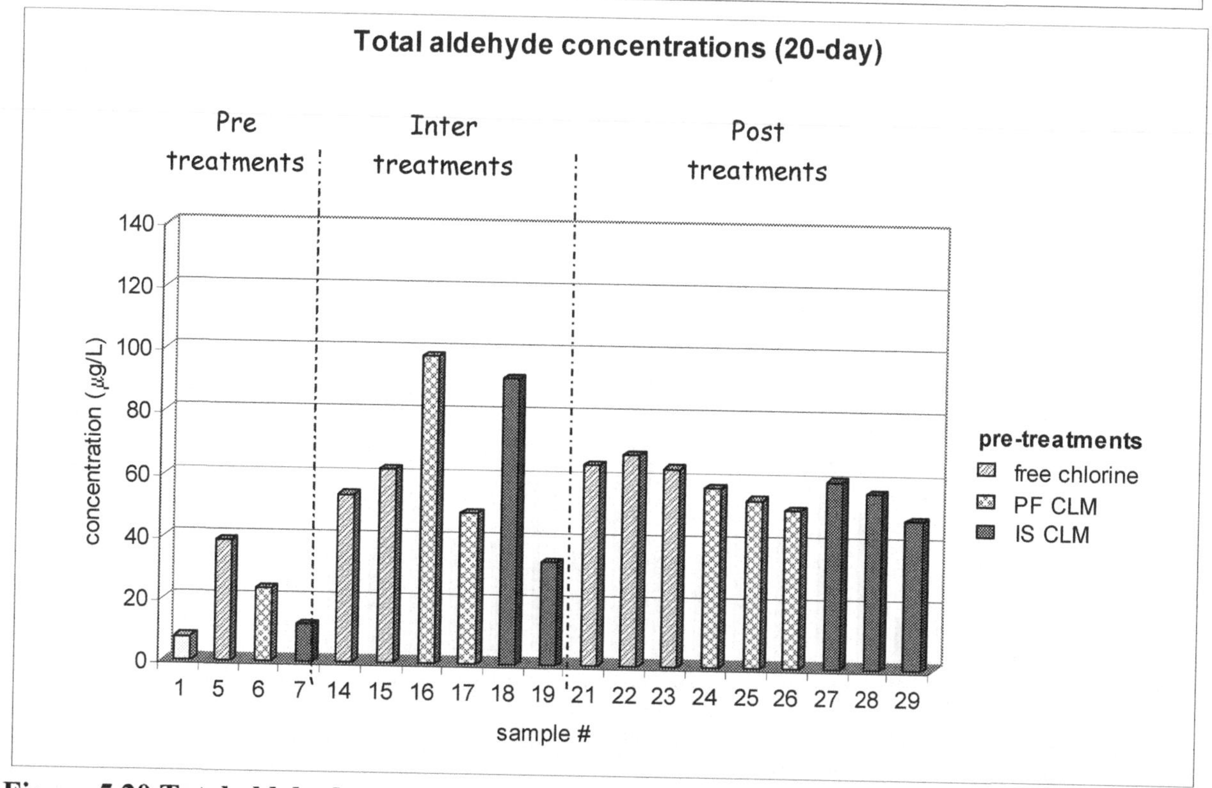

Figure 5.20 Total aldehyde results for 8-day and 20-day pretreatment tests – DBP Test No. 2 – July 16, 2001

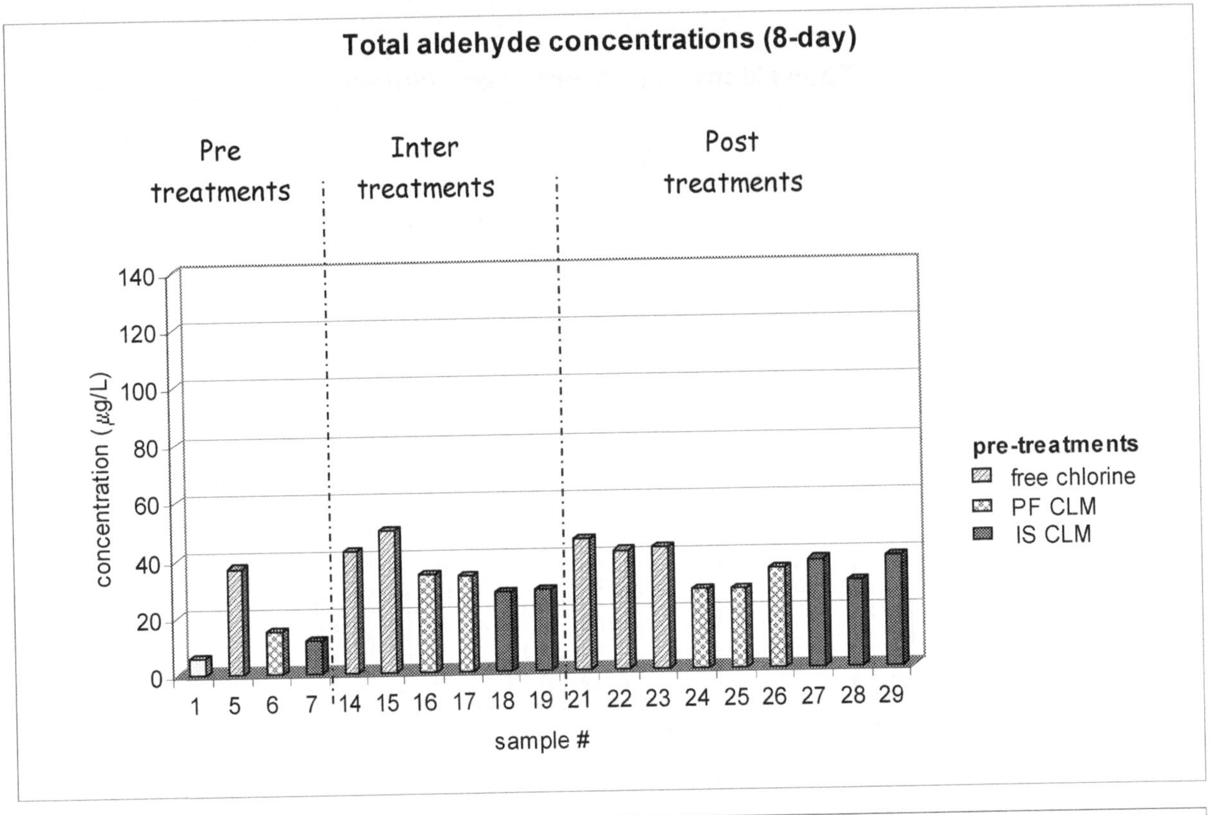

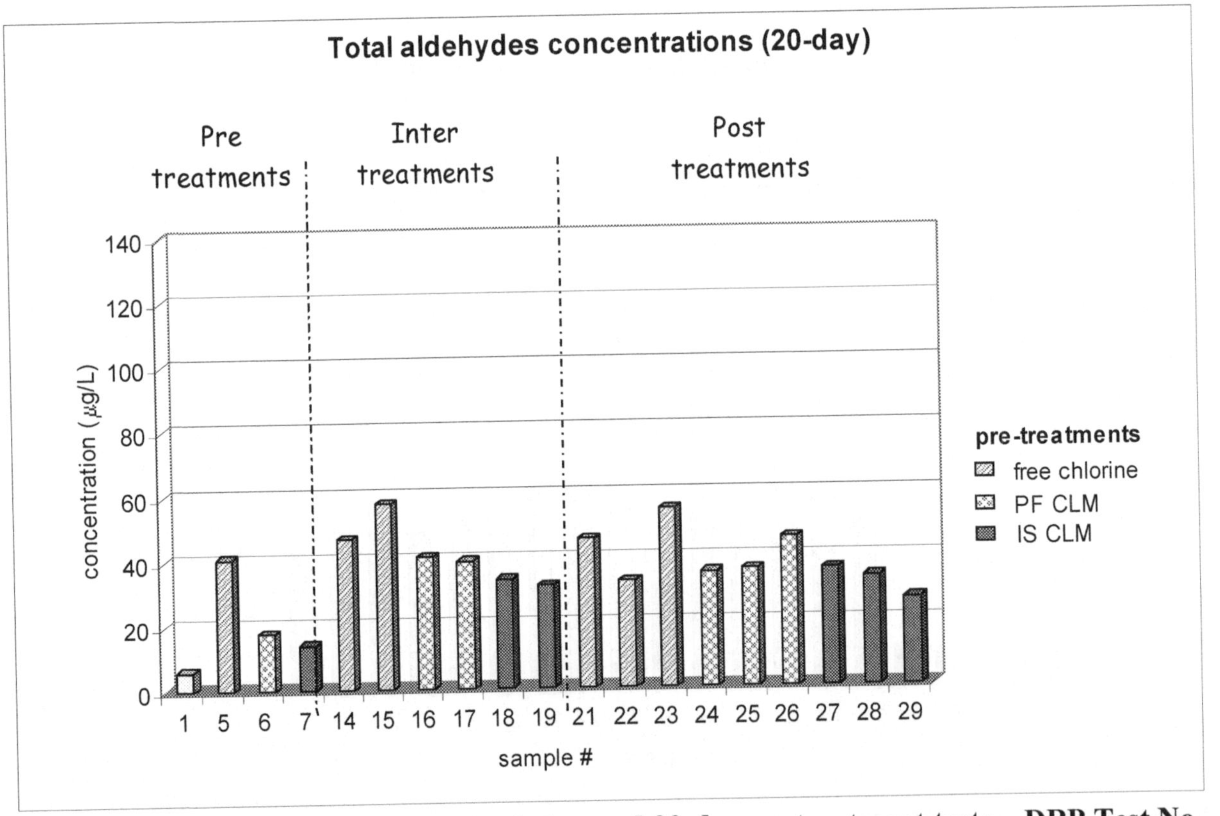

Figure 5.21 Total aldehyde results for 8-day and 20-day pretreatment tests – DBP Test No. 3 – October 10, 2001

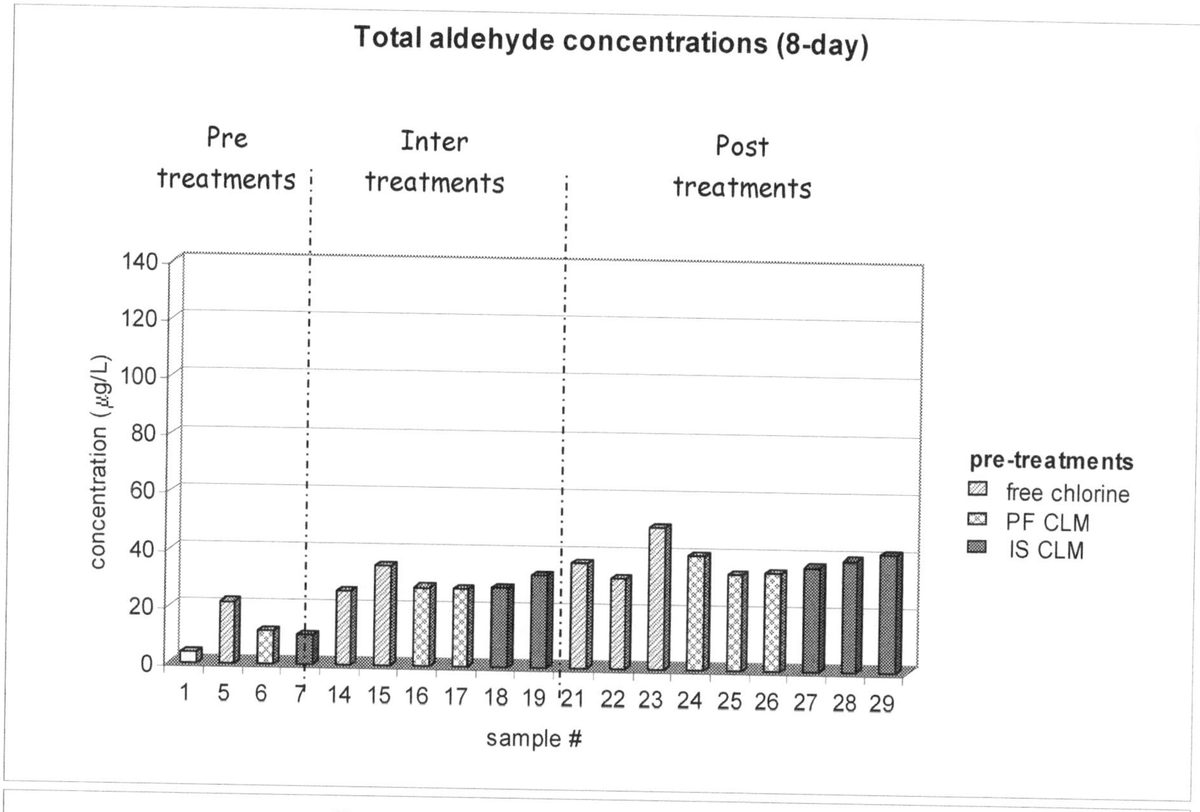

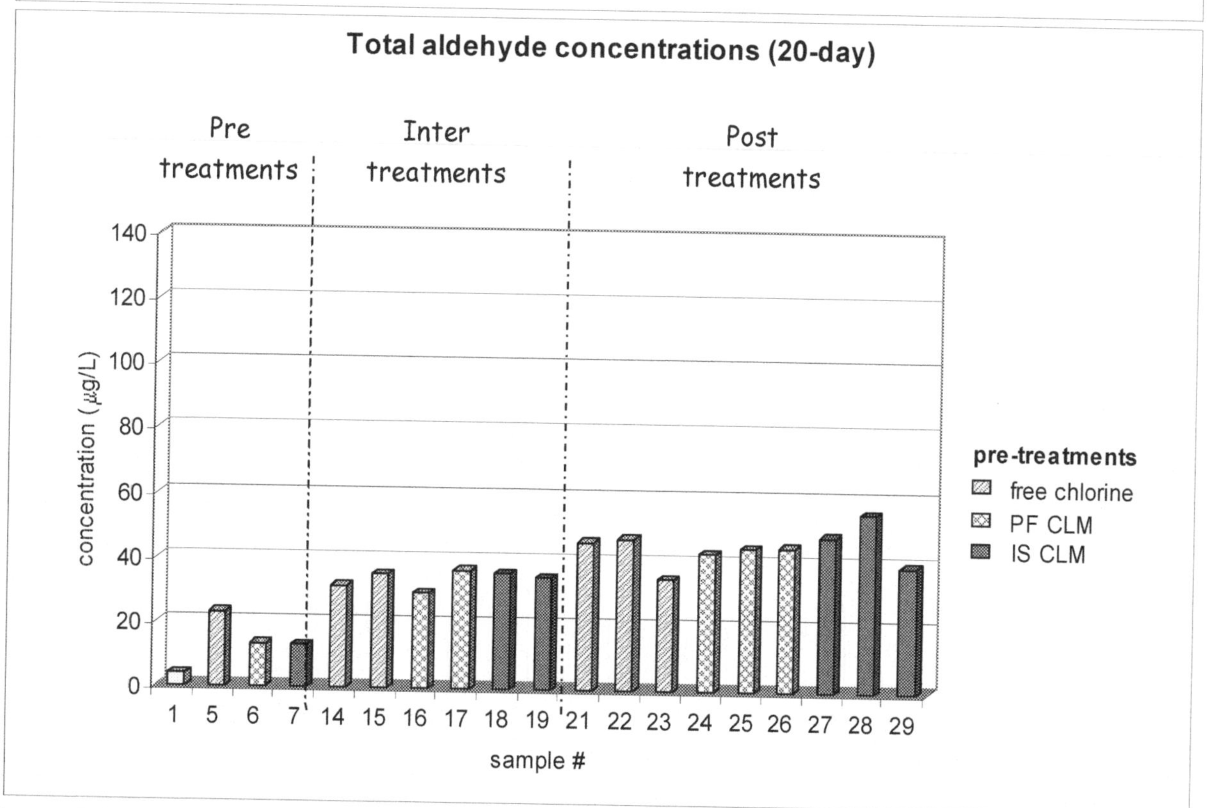

Figure 5.22 Total aldehyde results for 8-day and 20-day pretreatment tests – DBP Test No. 4 – November 26, 2001

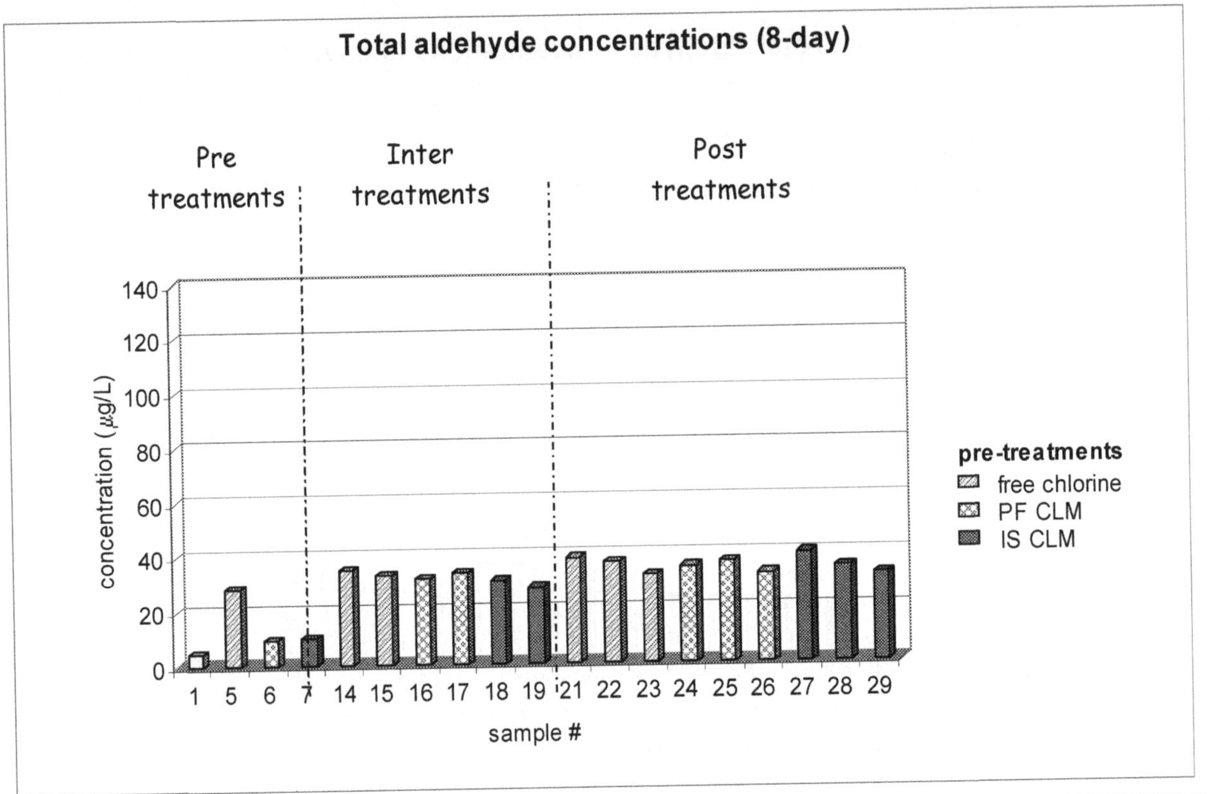

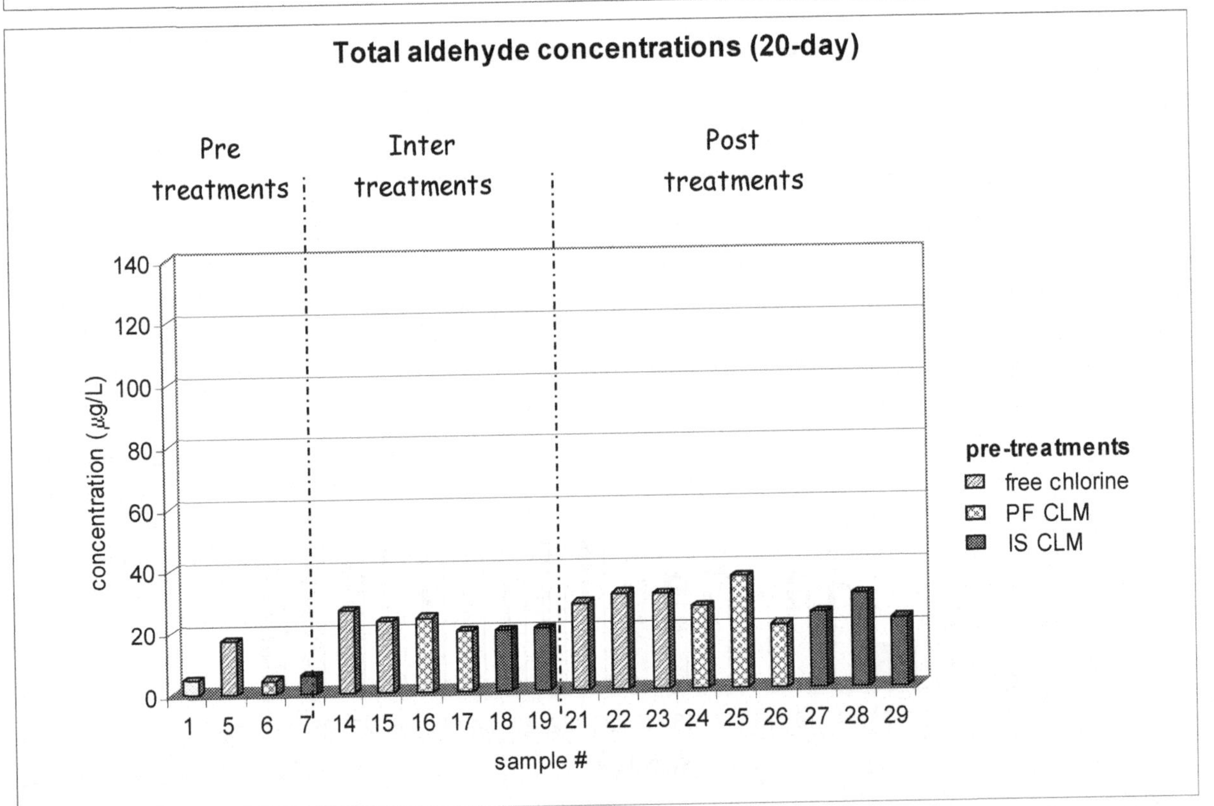

Figure 5.23 Total aldehyde results for 8-day and 20-day pretreatment tests – DBP Test No. 5 – January 30, 2002

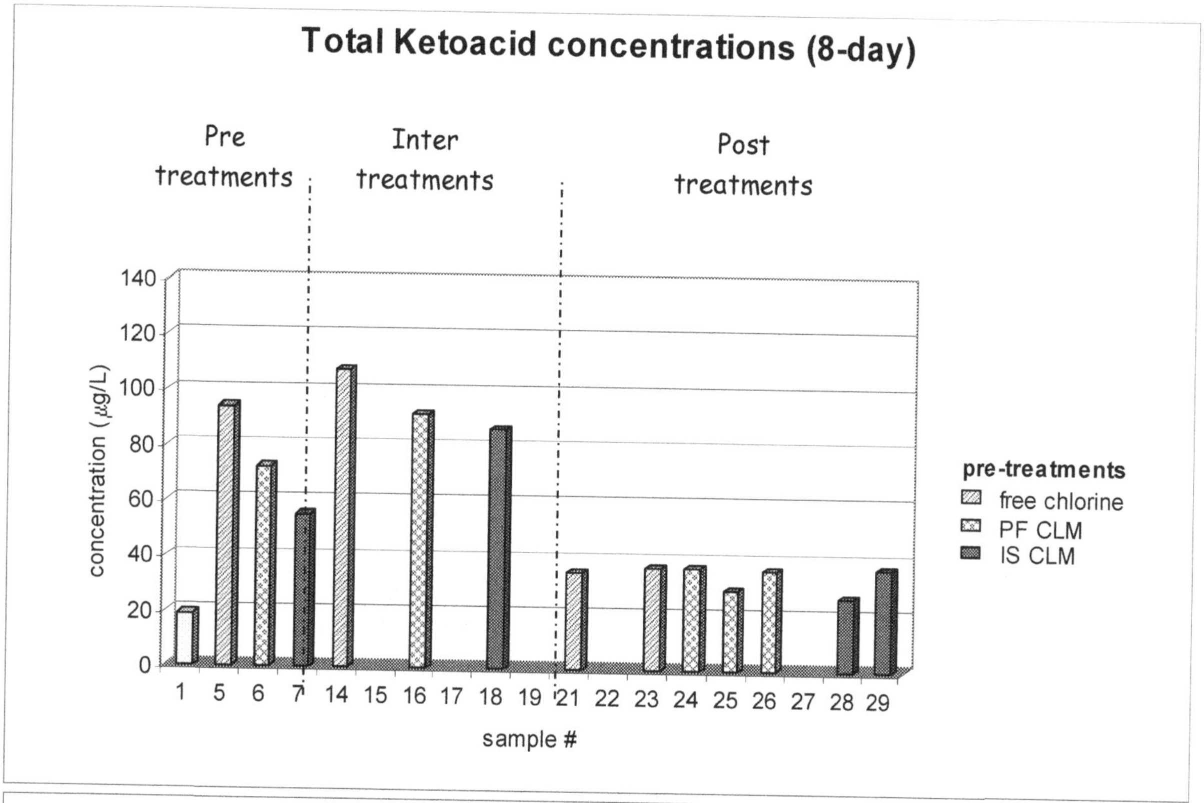

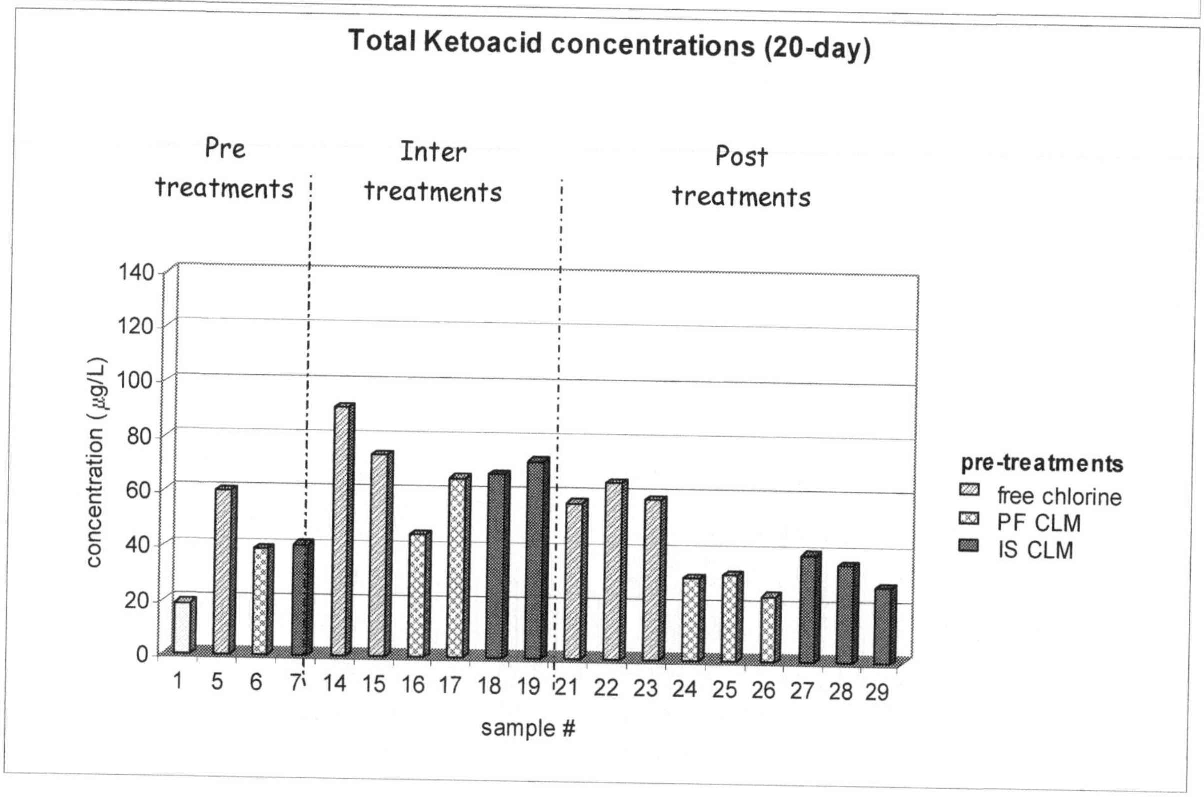

Figure 5.24 Total ketoacid results for 8-day and 20-day pretreatment tests – DBP Test No. 1 – June 8, 2001

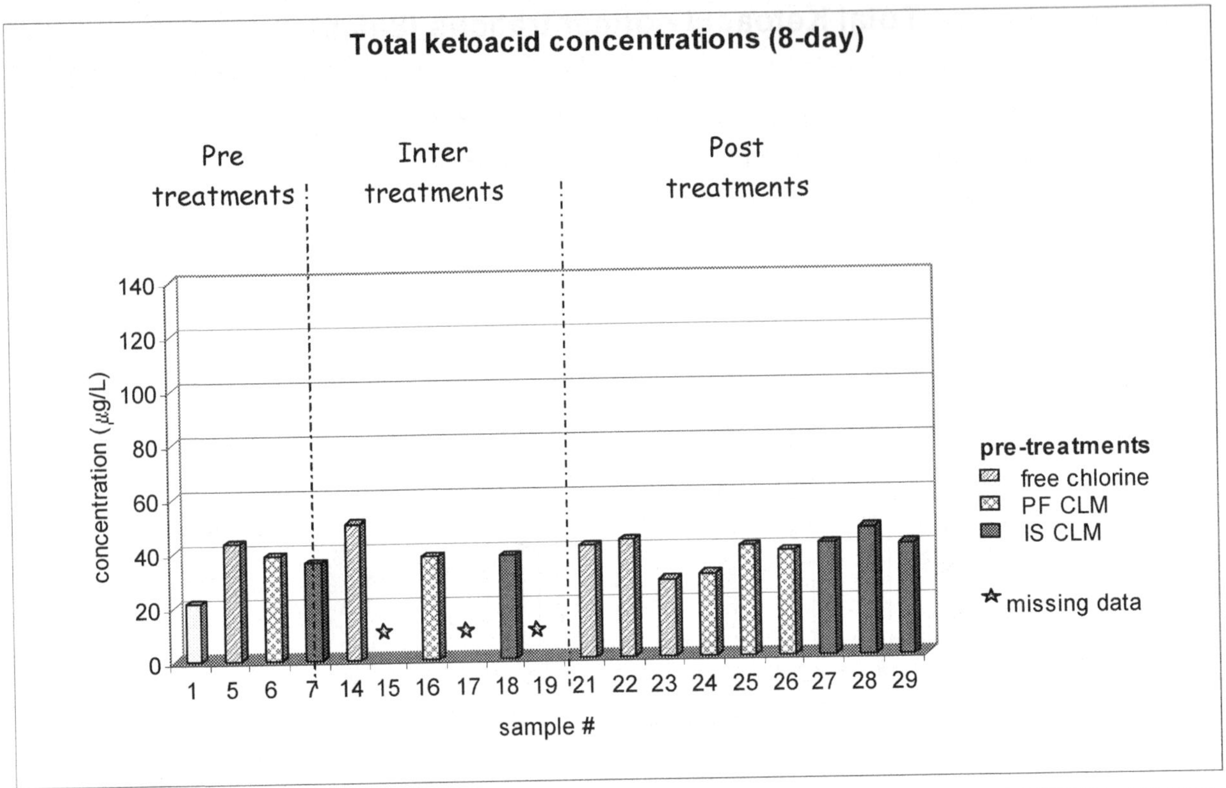

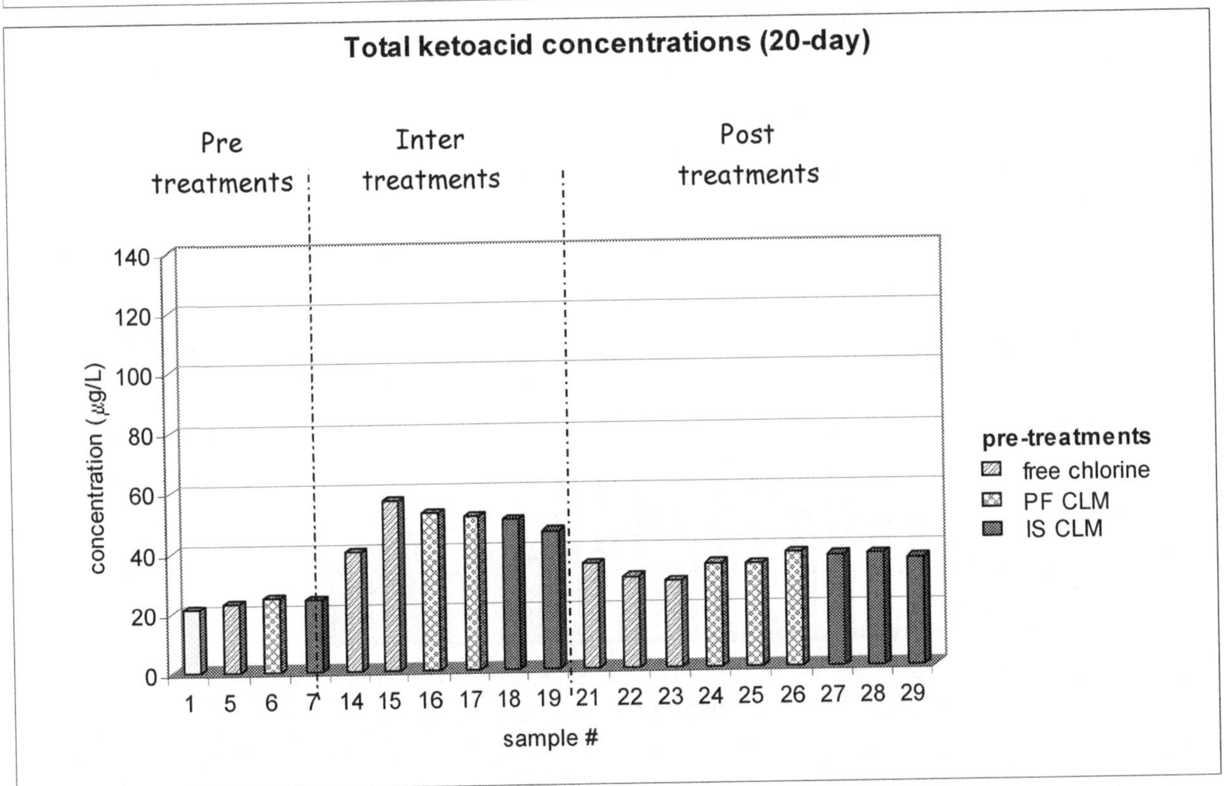

Figure 5.25 Total ketoacid results for 8-day and 20-day pretreatment tests – DBP Test No. 2 – July 16, 2001

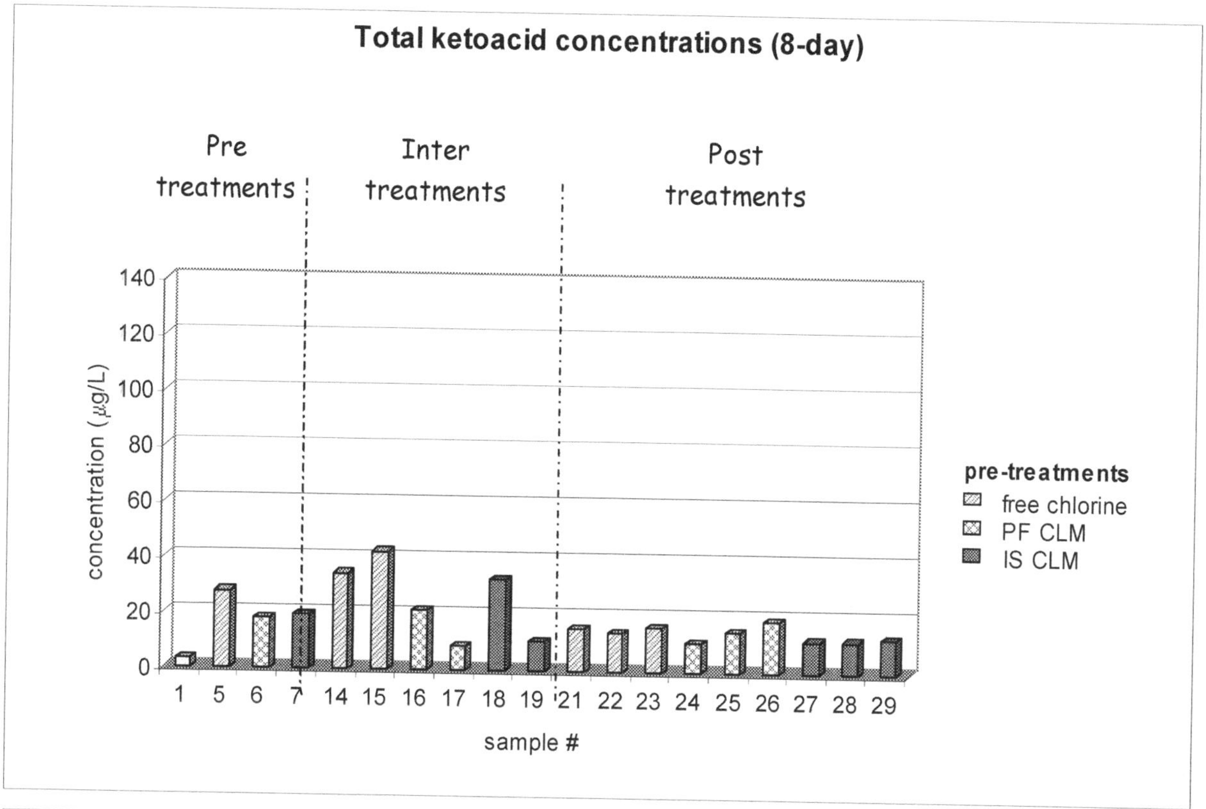

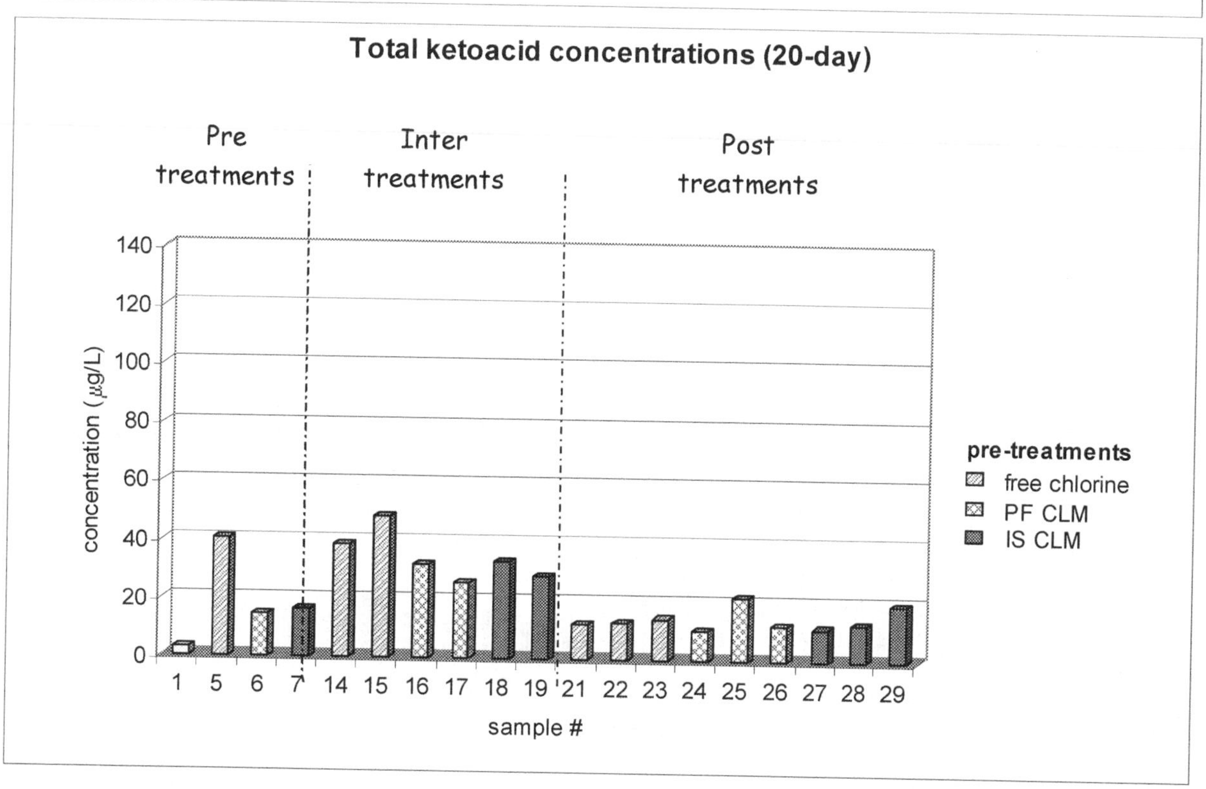

Figure 5.26 Total ketoacid results for 8-day and 20-day pretreatment tests – DBP Test No. 3 – October 10, 2001

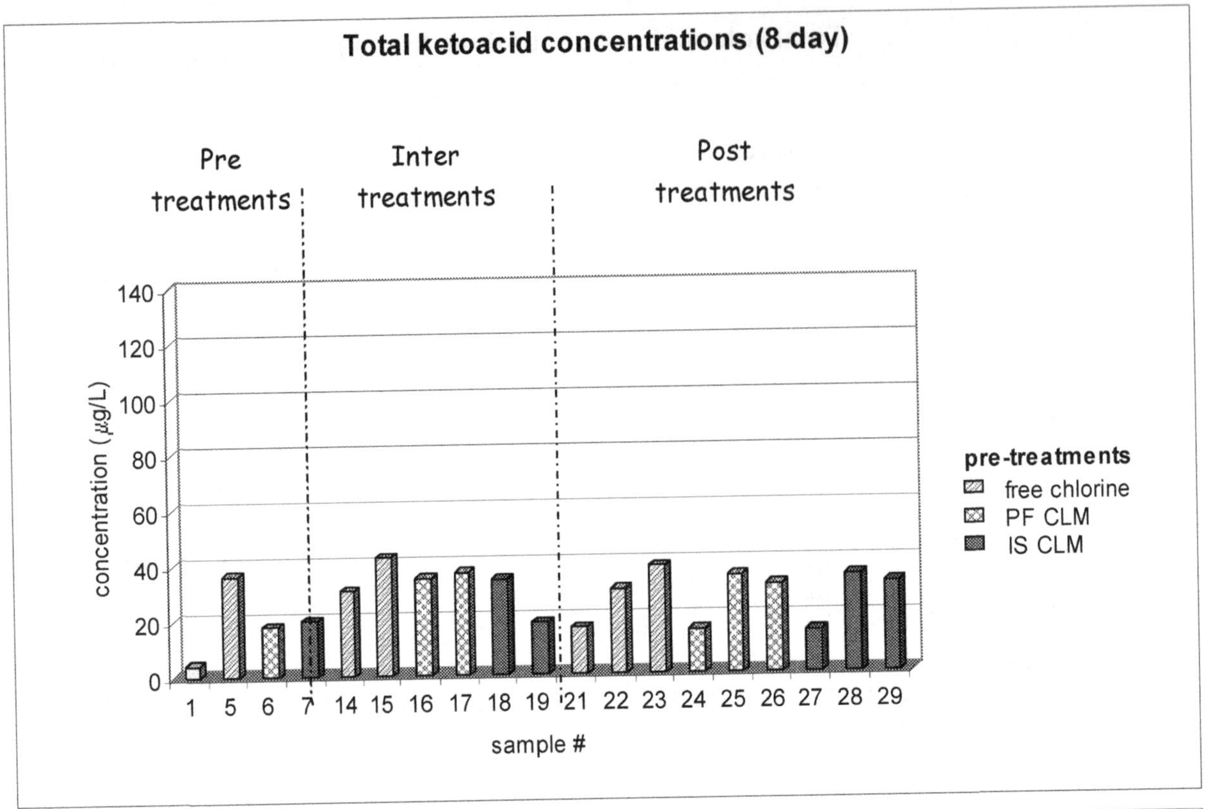

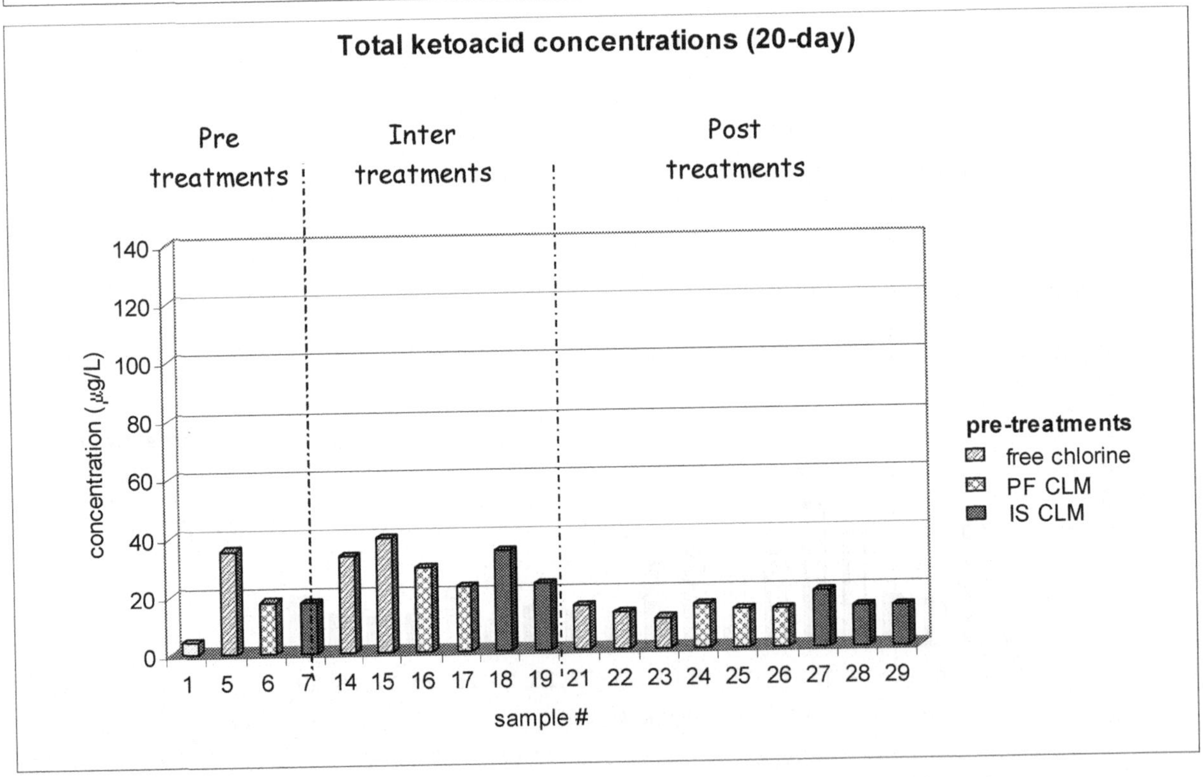

Figure 5.27 Total ketoacid results for 8-day and 20-day pretreatment tests – DBP Test No. 4 – November 26, 2001

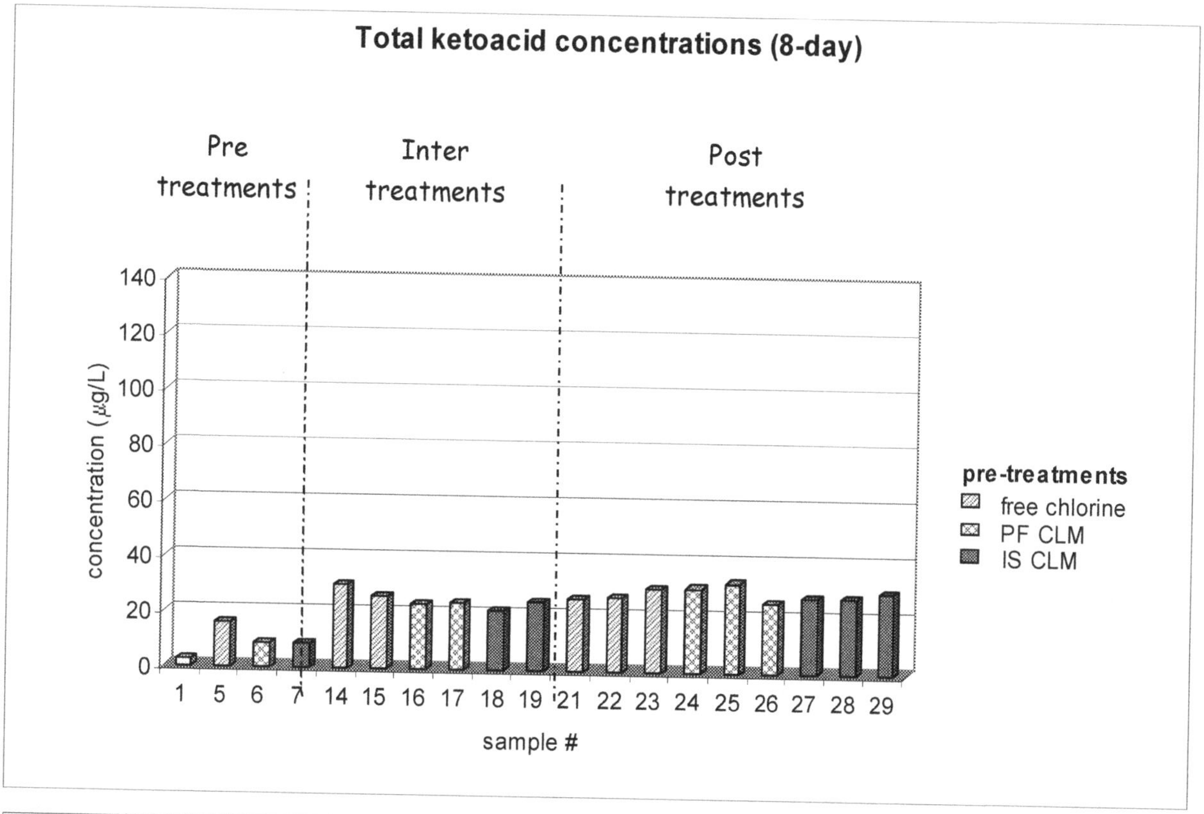

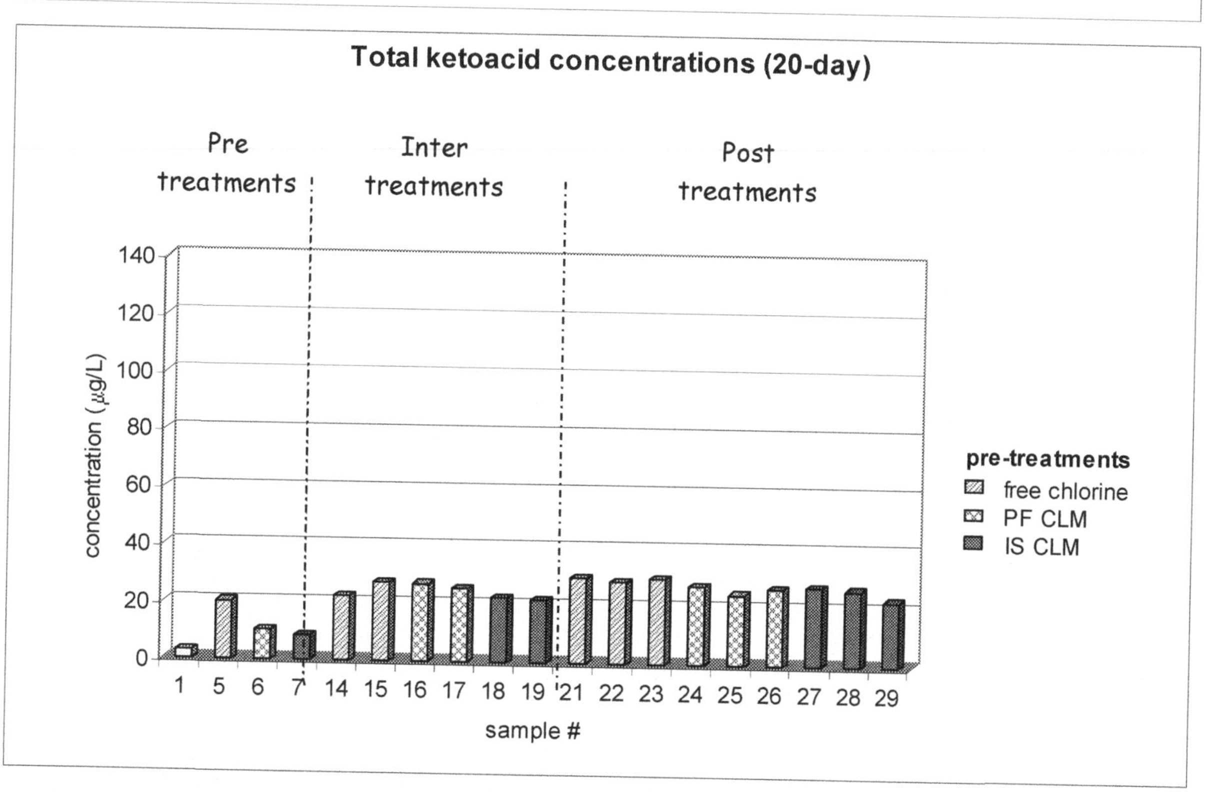

Figure 5.28 Total ketoacid results for 8-day and 20-day pretreatment tests – DBP Test No. 5 – January 30, 2002

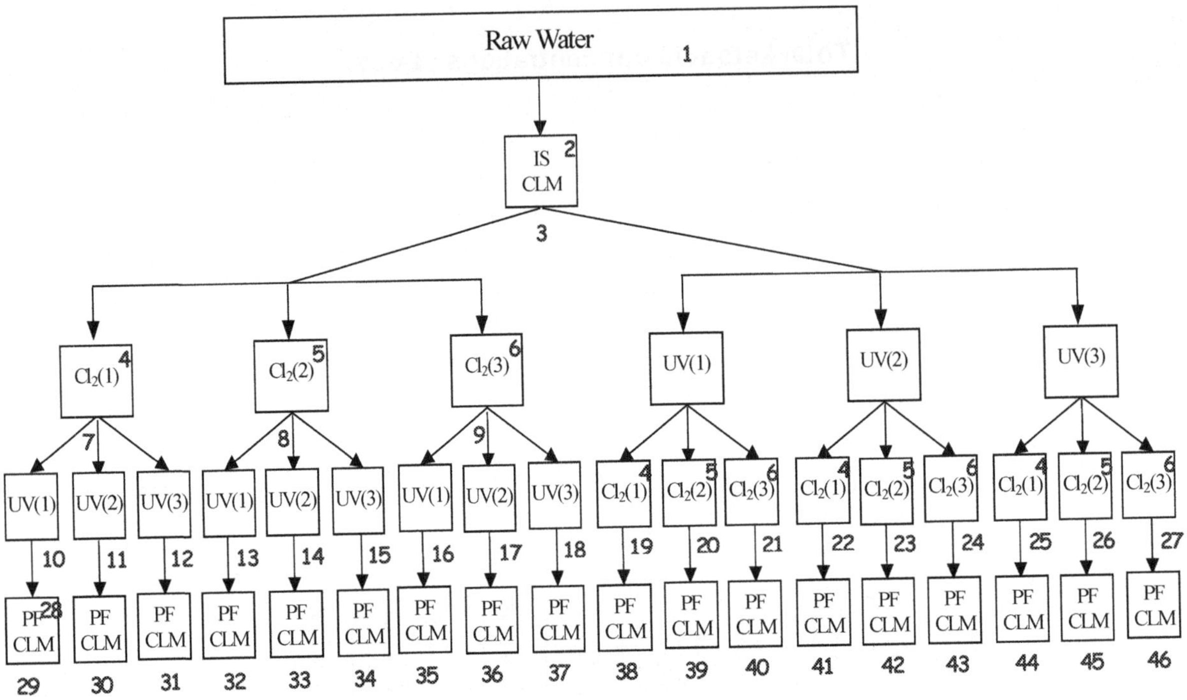

Figure 5.29 Sample number key for DBP Test No. 6

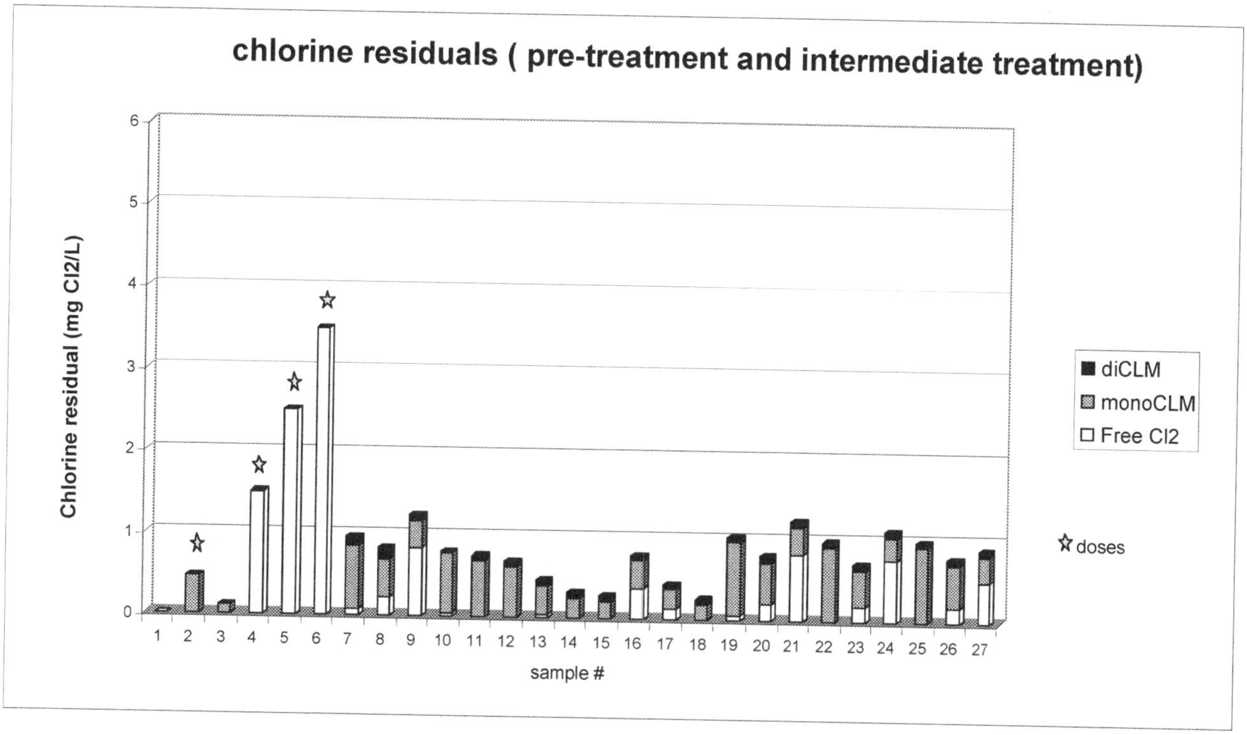

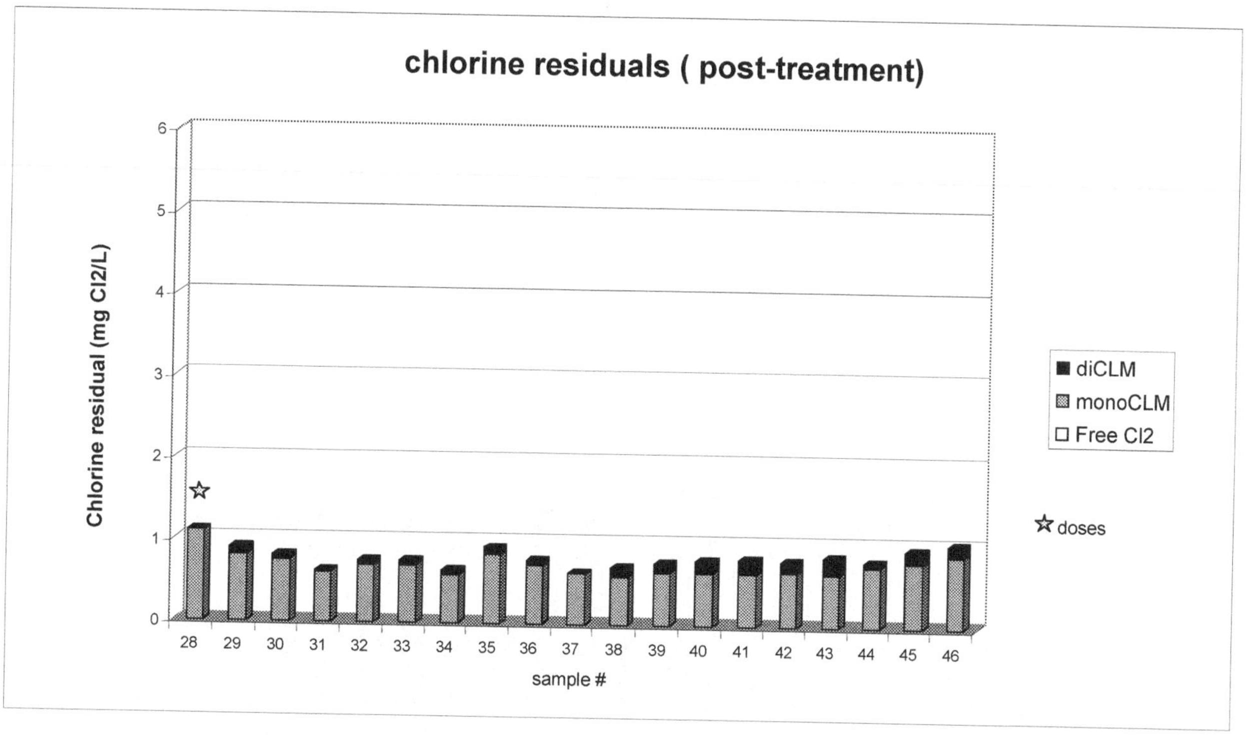

Figure 5.30 Chlorine dose and residual data – DBP Test No. 6

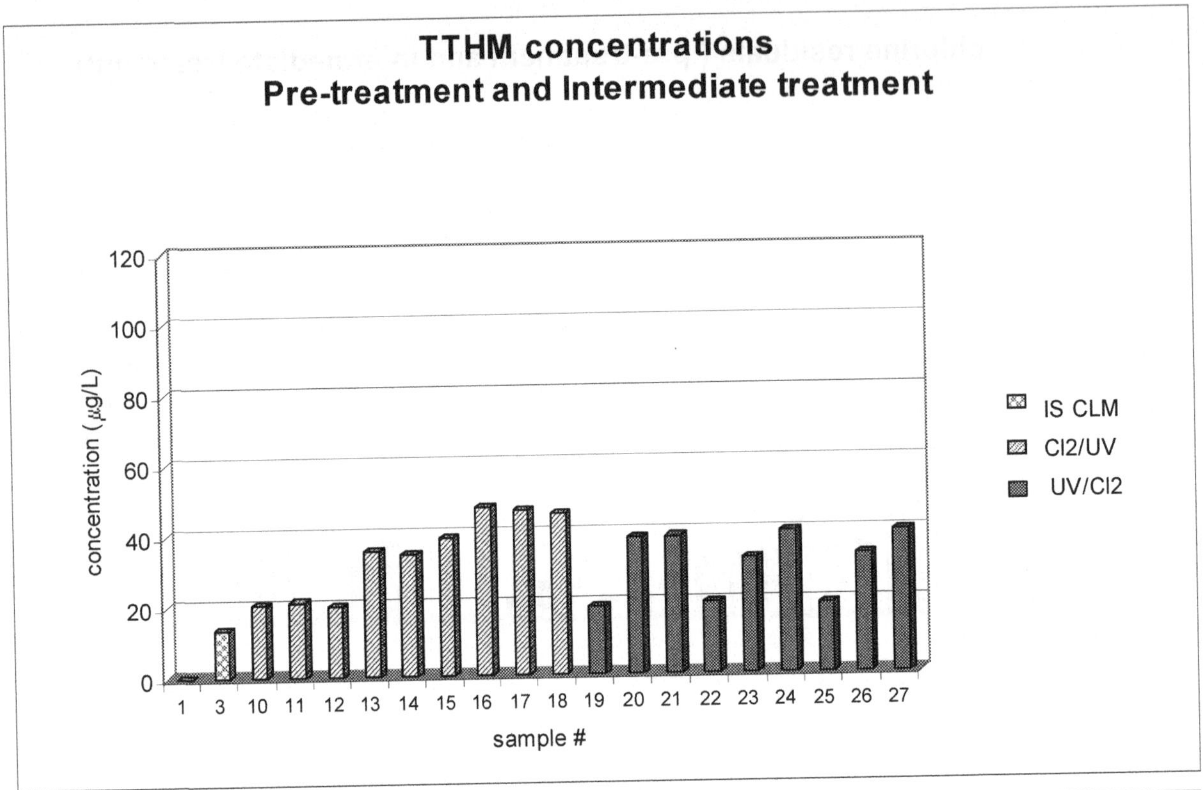

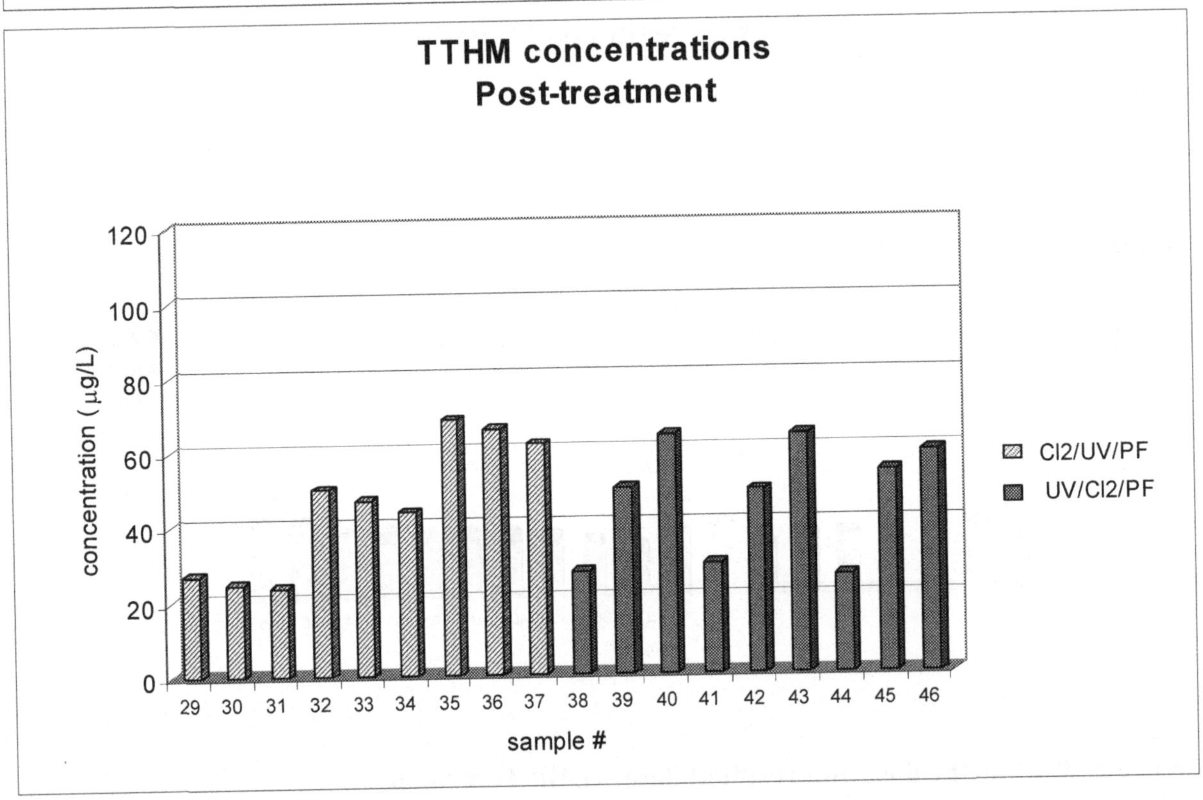

Figure 5.31 TTHM results for DBP Test No. 6

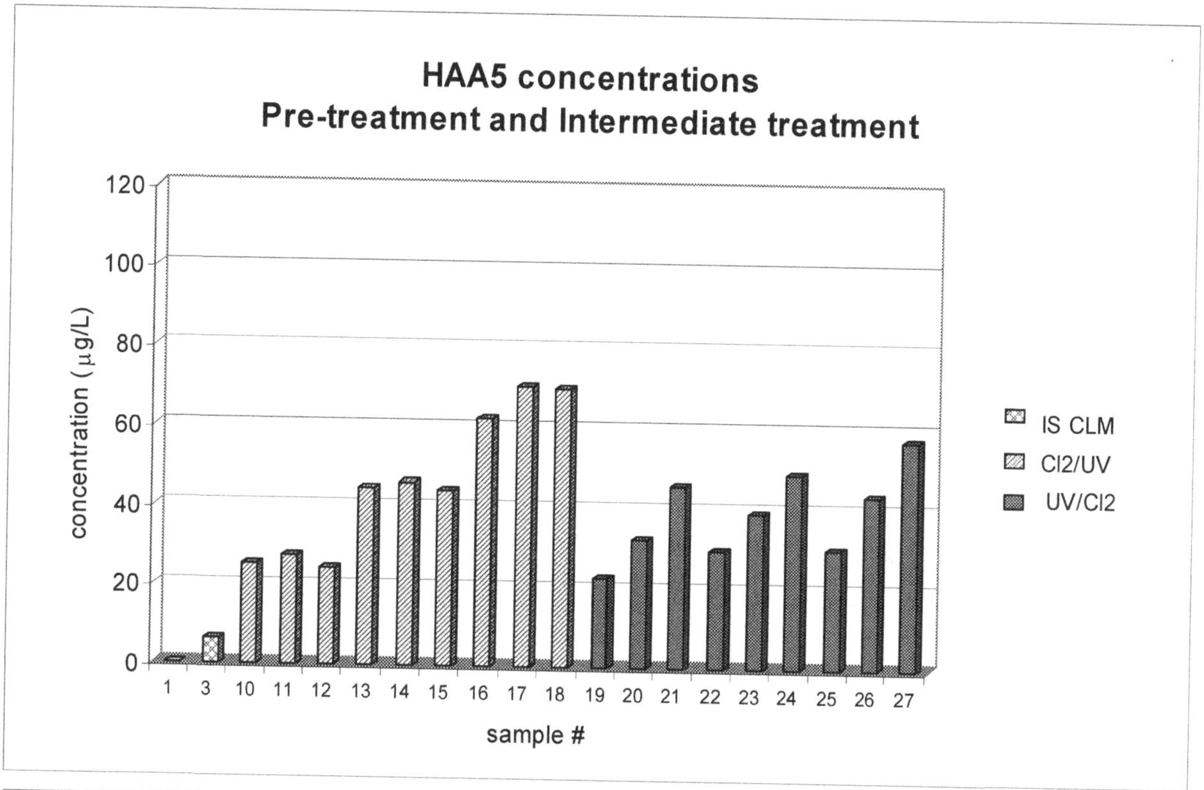

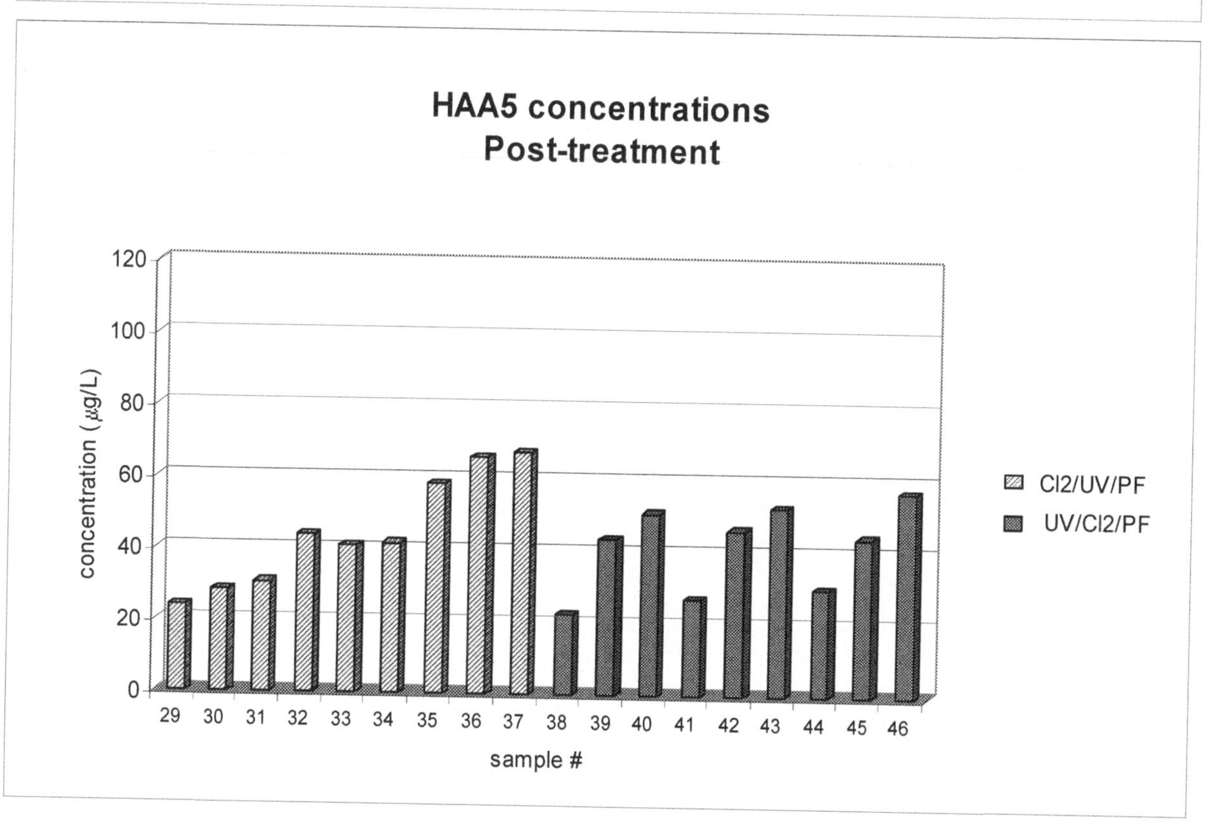

Figure 5.32 HAA results for DBP Test No. 6

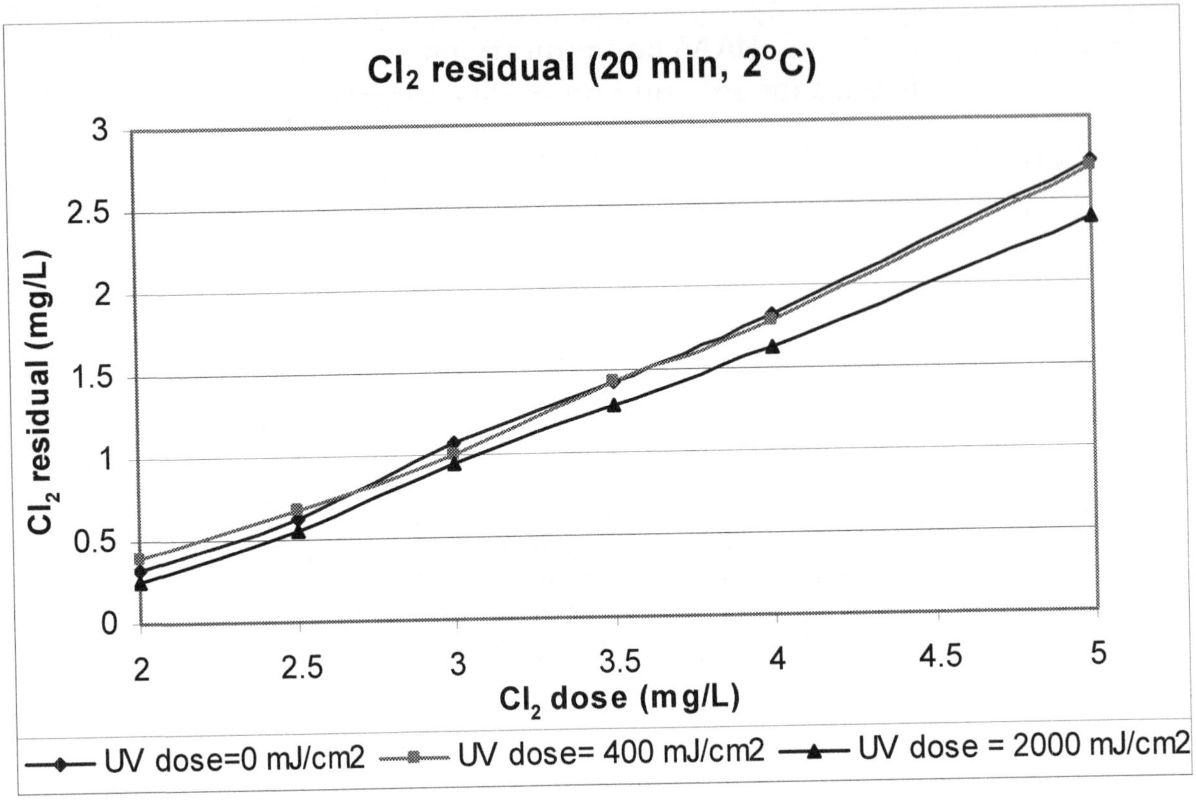

Figure 5.33 UV/chlorine residual data – DBP Test No. 6

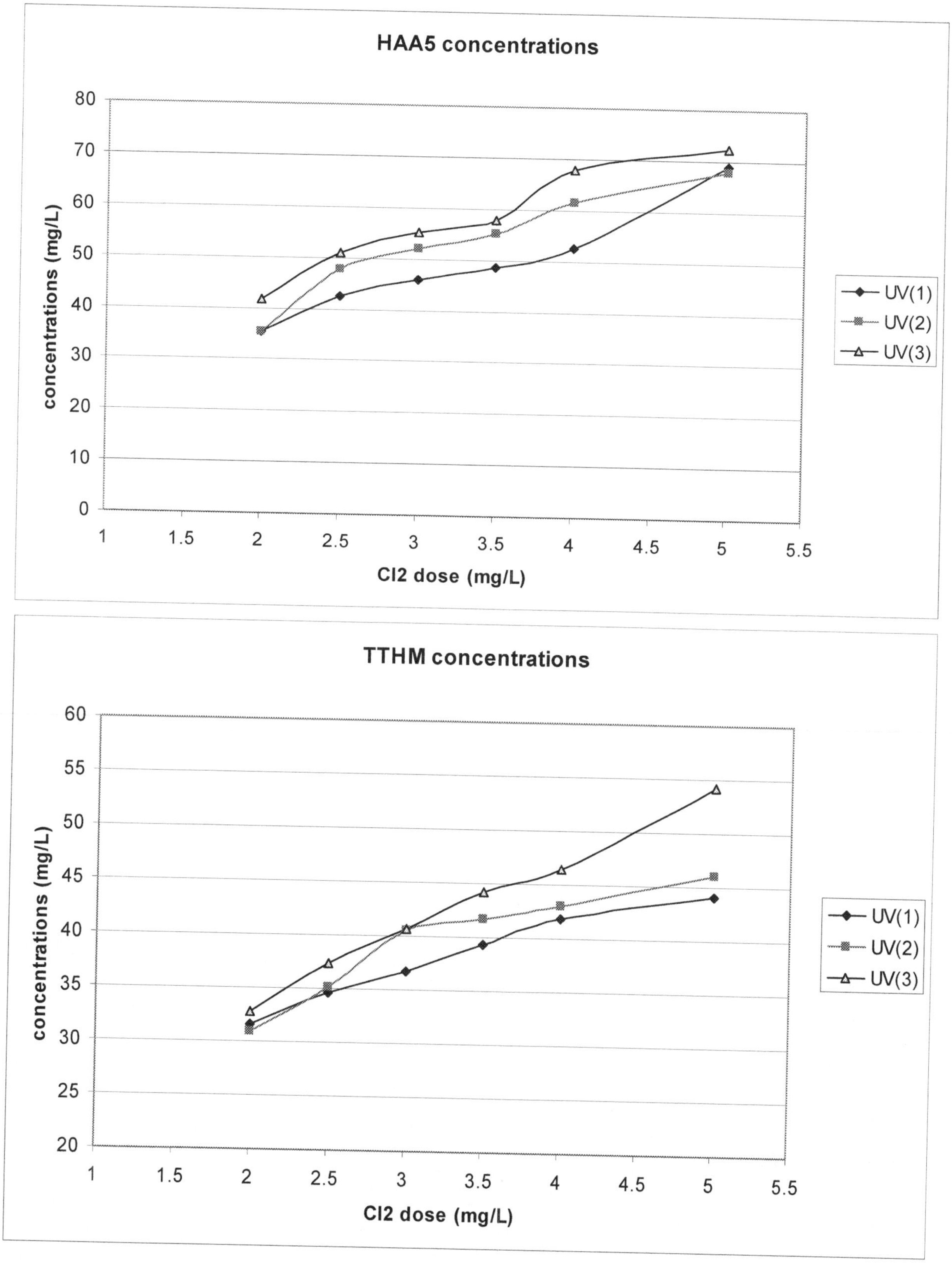

Figure 5.34 TTHM and HAA5 concentrations versus UV dose – DBP Test No. 6

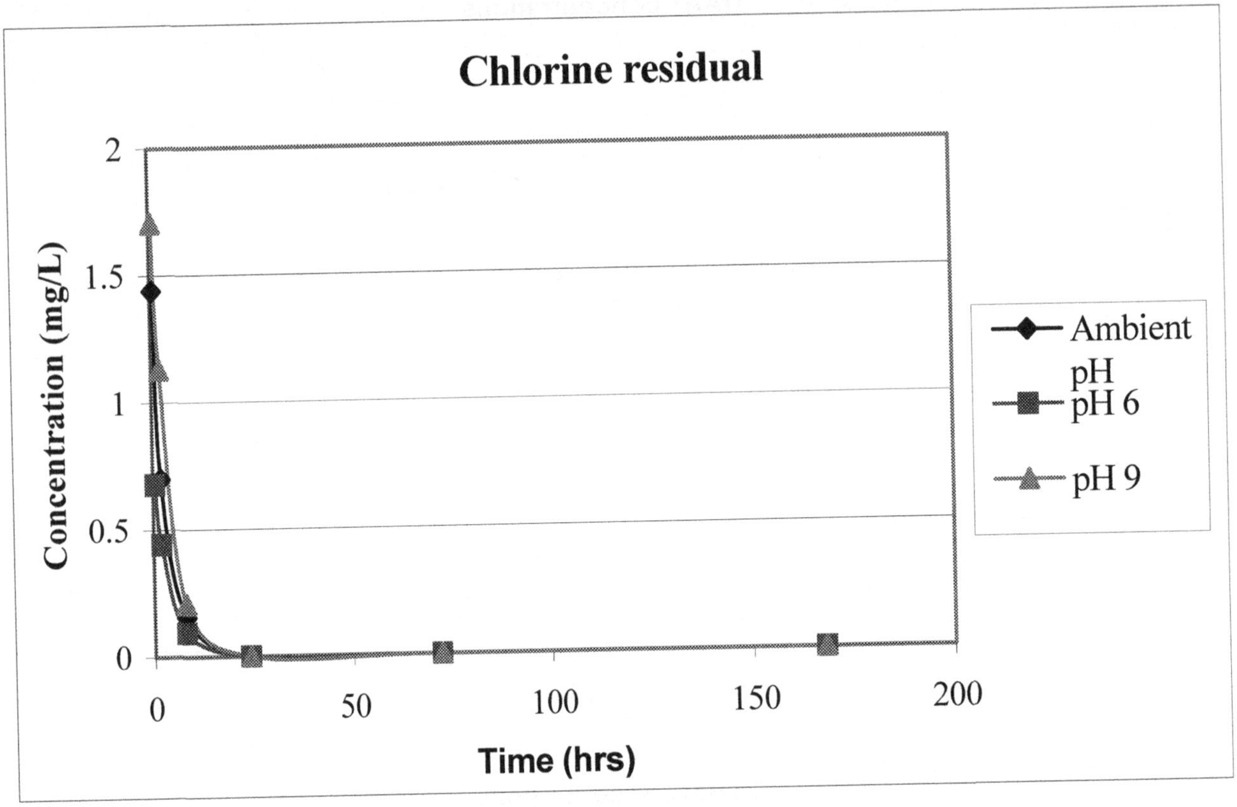

Figure 5.35 pH effect on chlorine residual – chlorine dose = 3.4 mg/L (Ambient pH = 7.48)

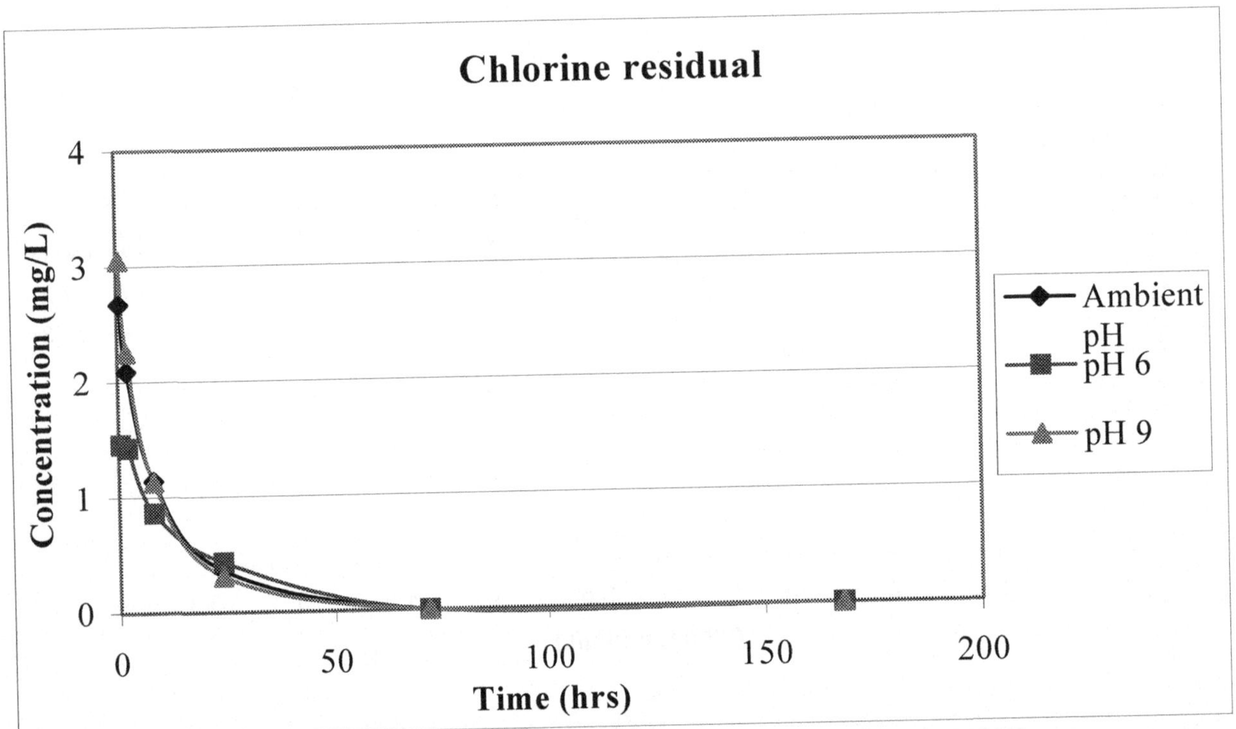

Figure 5.36 pH effect on chlorine residual – chlorine dose = 4.8 mg/L (Ambient pH = 7.48)

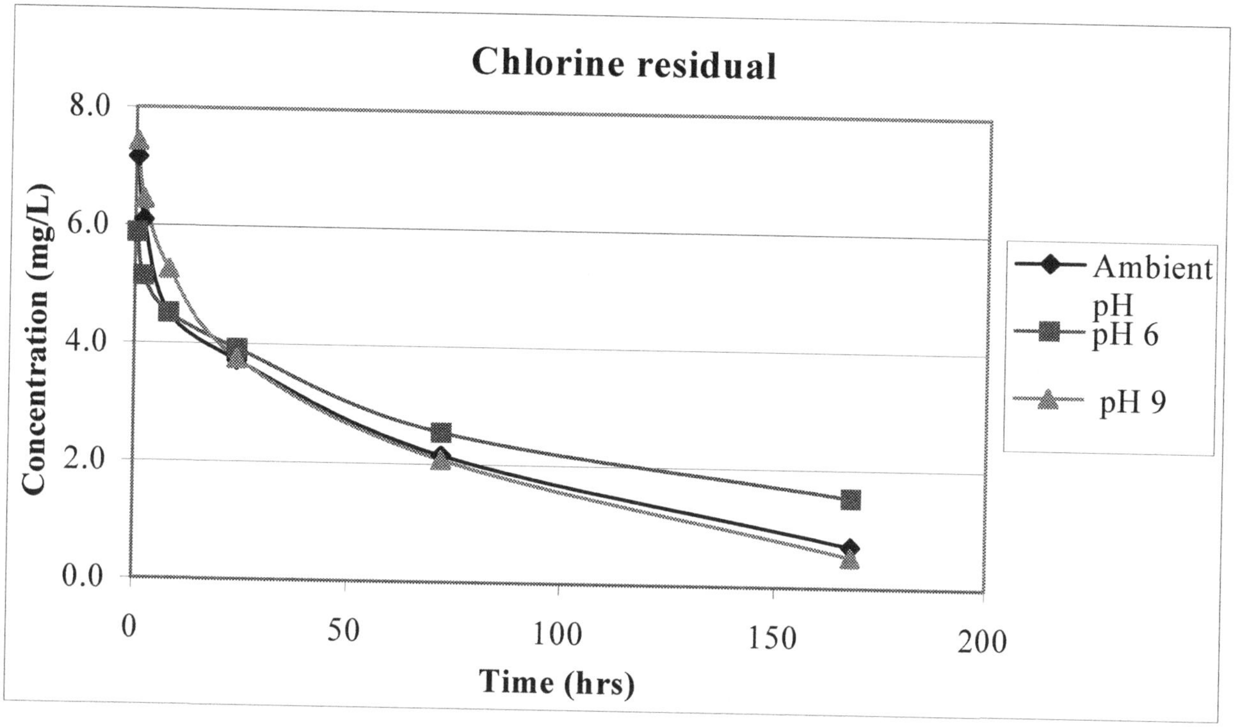

Figure 5.37 pH effect on chlorine residual – chlorine dose = 9.6 mg/L (Ambient pH = 7.48)

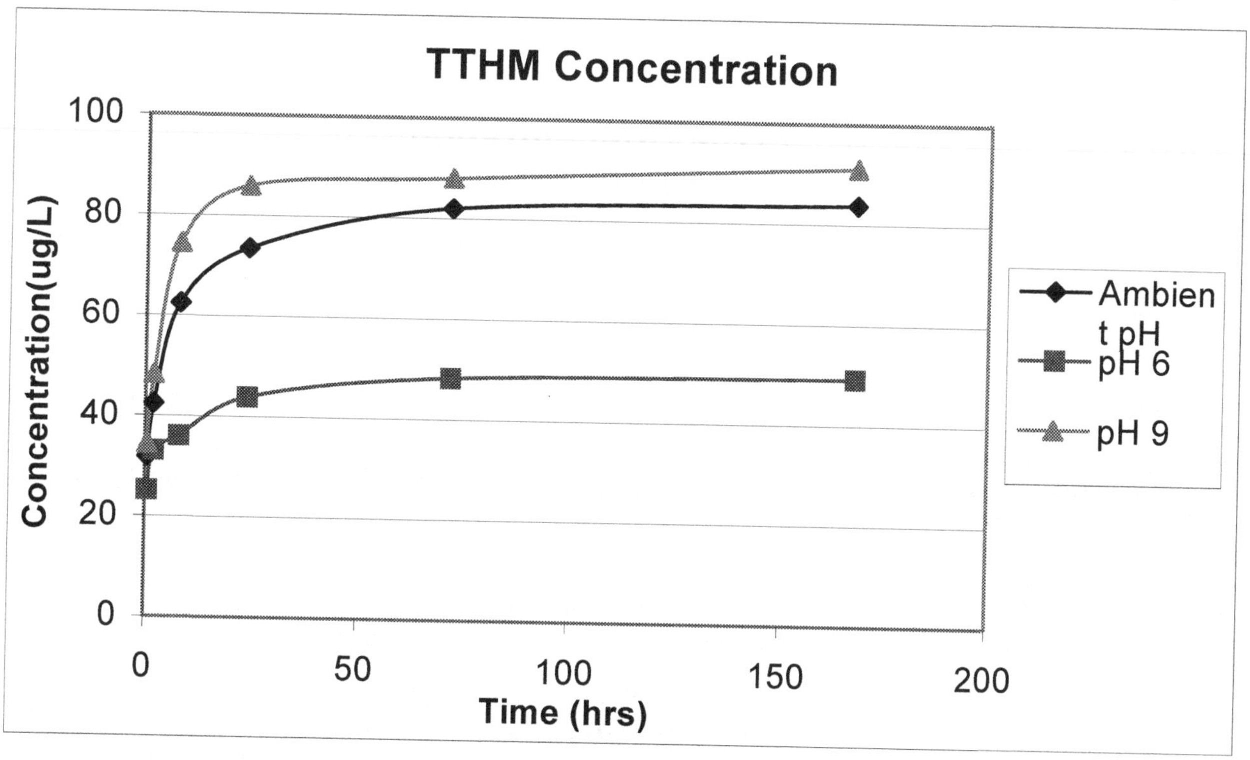

Figure 5.38 Effect of pH on TTHM concentration – chlorine dose 3.4 mg/L (Ambient pH = 7.48)

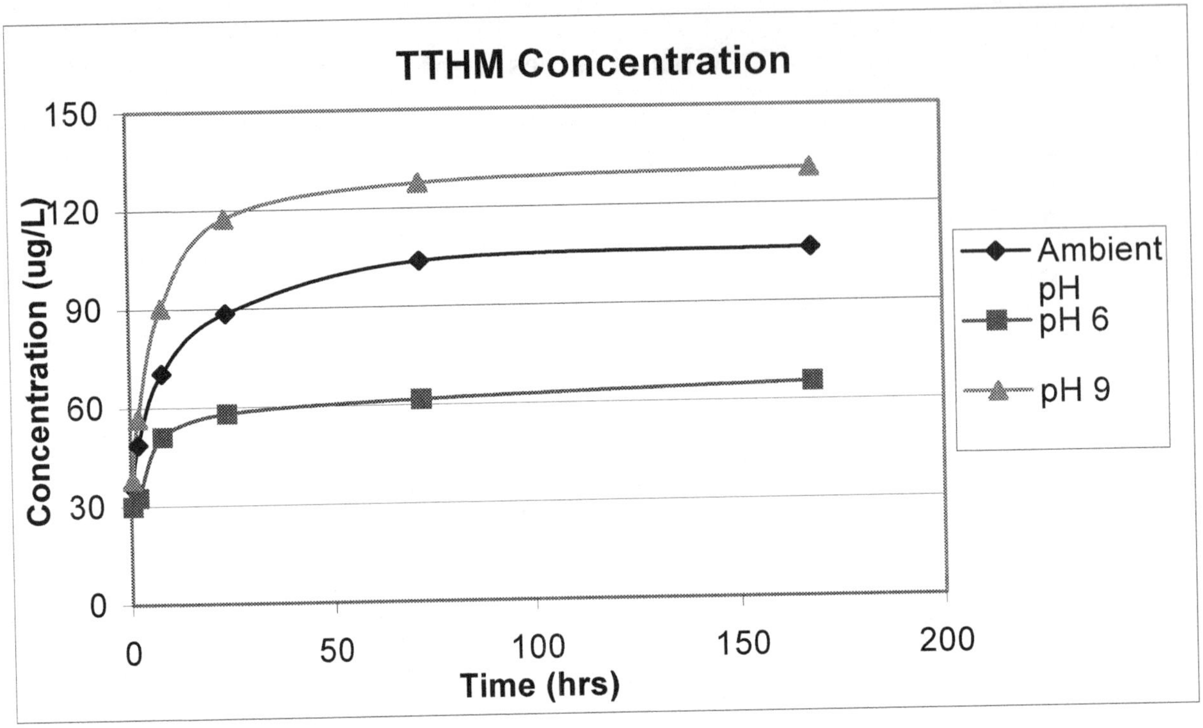

Figure 5.39 Effect of pH on TTHM concentration – chlorine dose 4.8 mg/L (Ambient pH = 7.48)

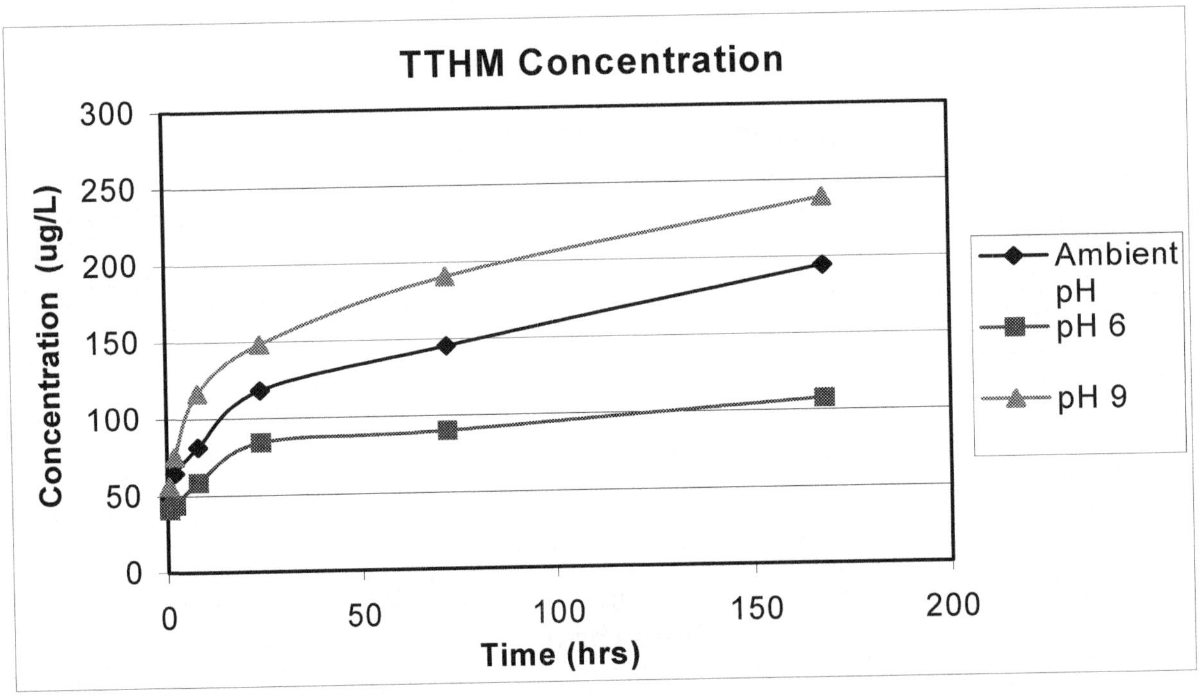

Figure 5.40 Effect of pH on TTHM concentration – chlorine dose 9.6 mg/L (Ambient pH = 7.48)

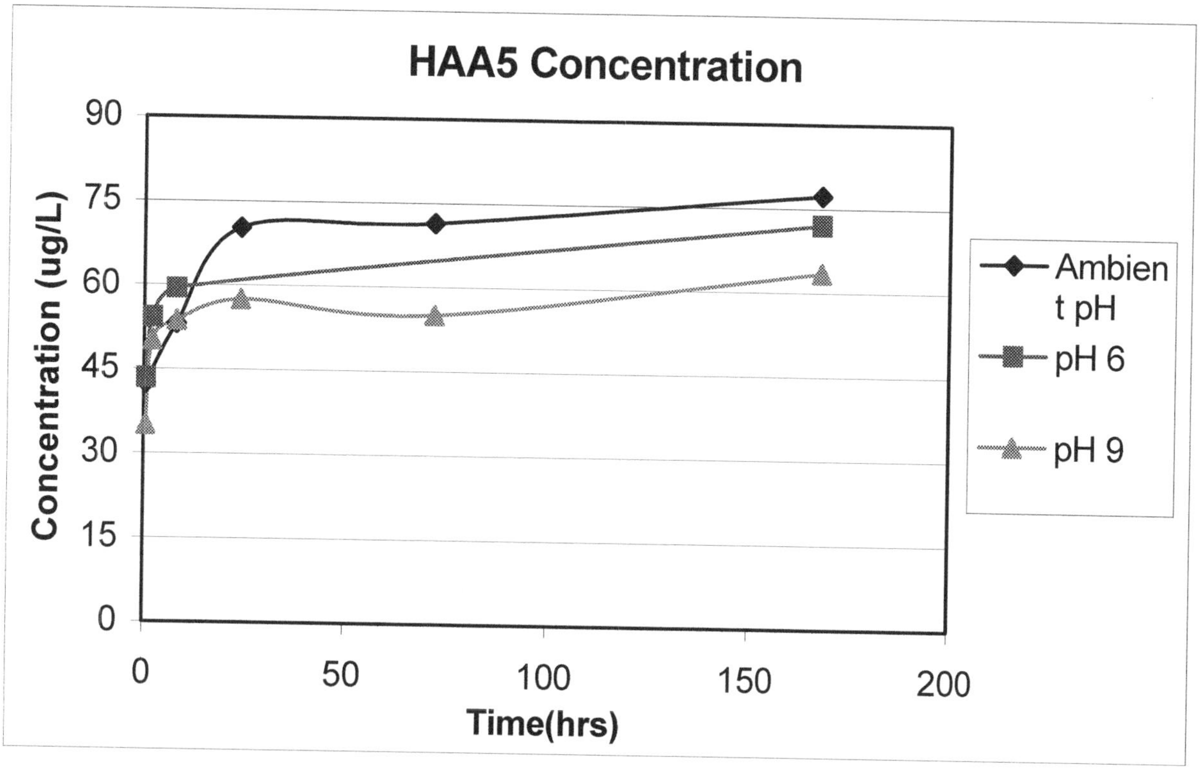

Figure 5.41 Effect of pH on HAA5 – chlorine dose 3.4 mg/L (Ambient pH = 7.48)

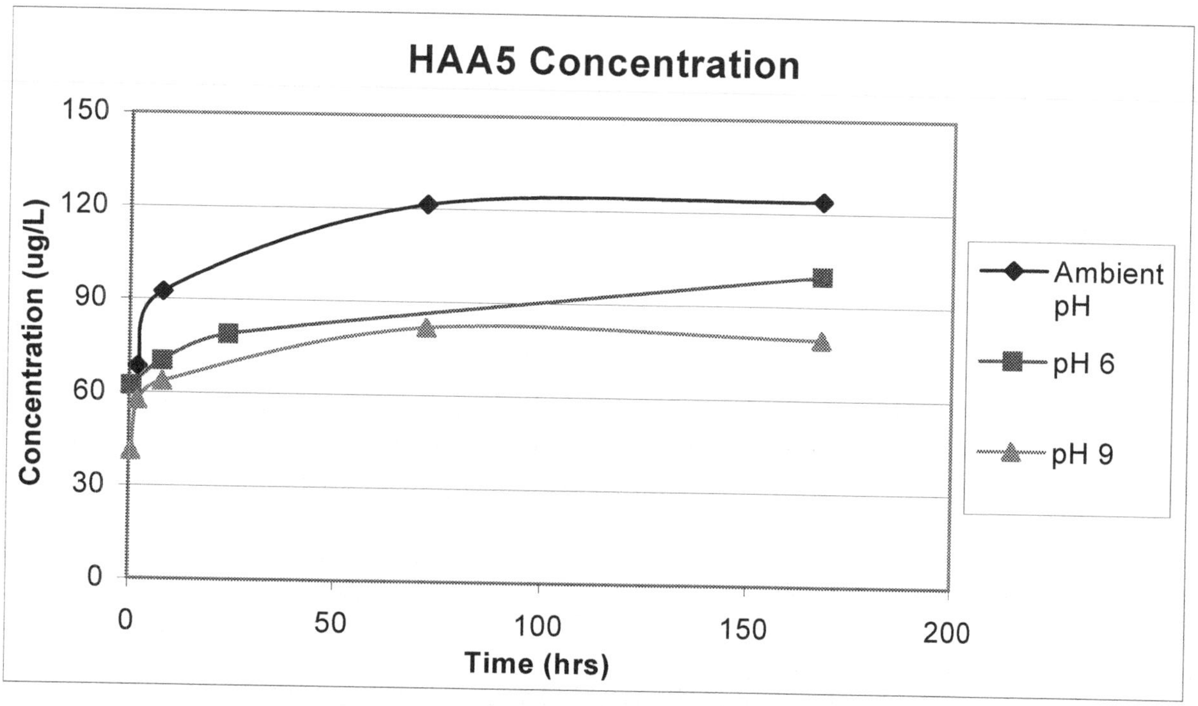

Figure 5.42 Effect of pH on HAA5 – chlorine dose 4.8 mg/L (Ambient pH = 7.48)

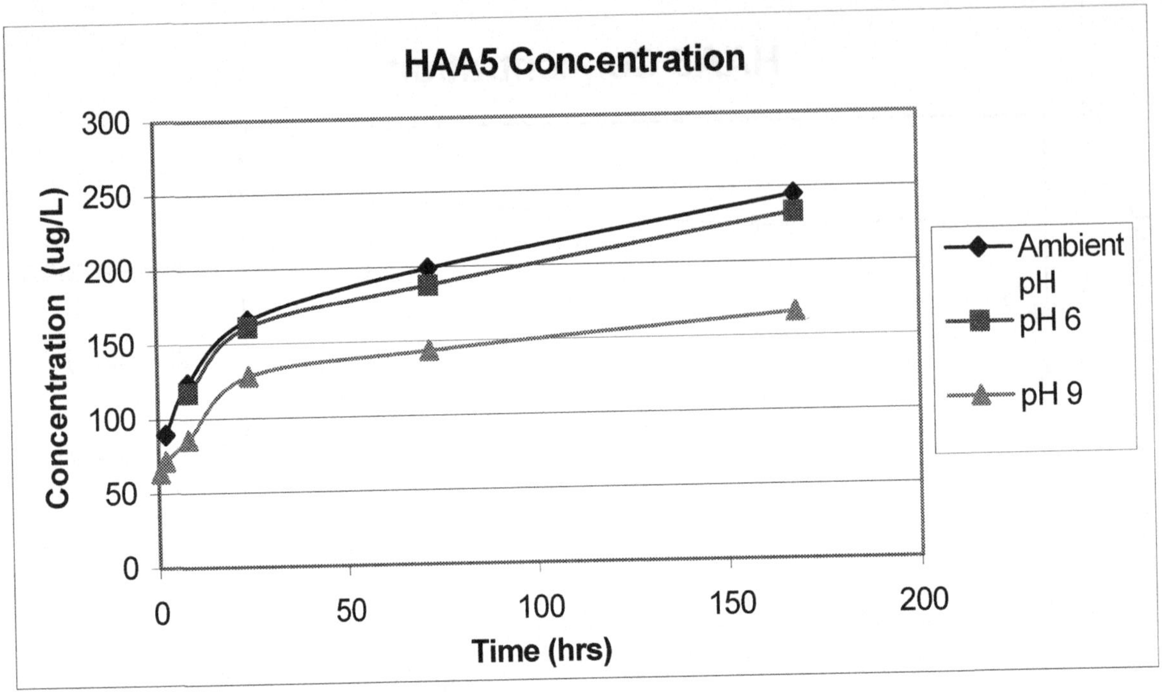

Figure 5.43 Effect of pH on HAA5 – chlorine dose 9.6 mg/L (Ambient pH = 7.48)

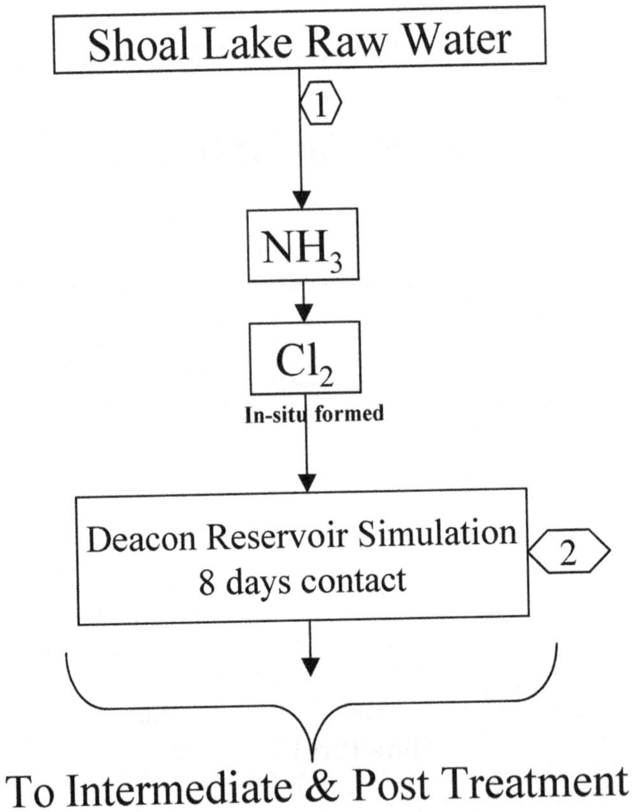

Figure 5.44 Pretreatment bench-scale experimental schematic – DBP Test No. 8 (hexagons show sample numbers)

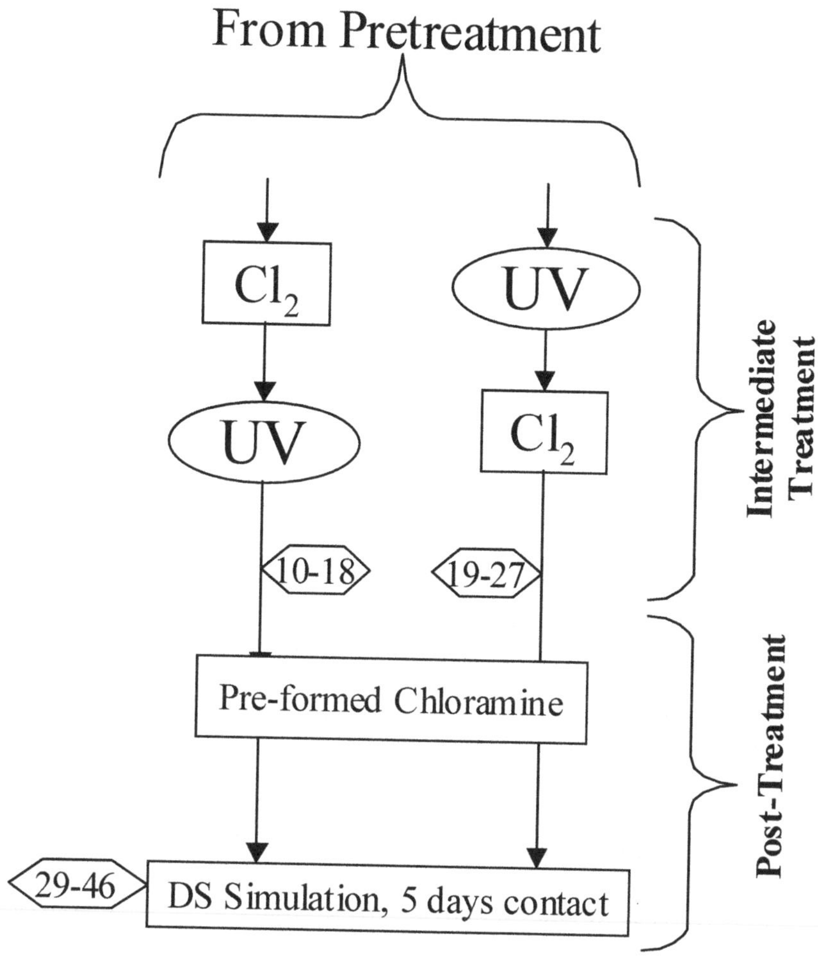

Figure 5.45 Post-treatment bench-scale experimental schematic (hexagons show sample numbers)

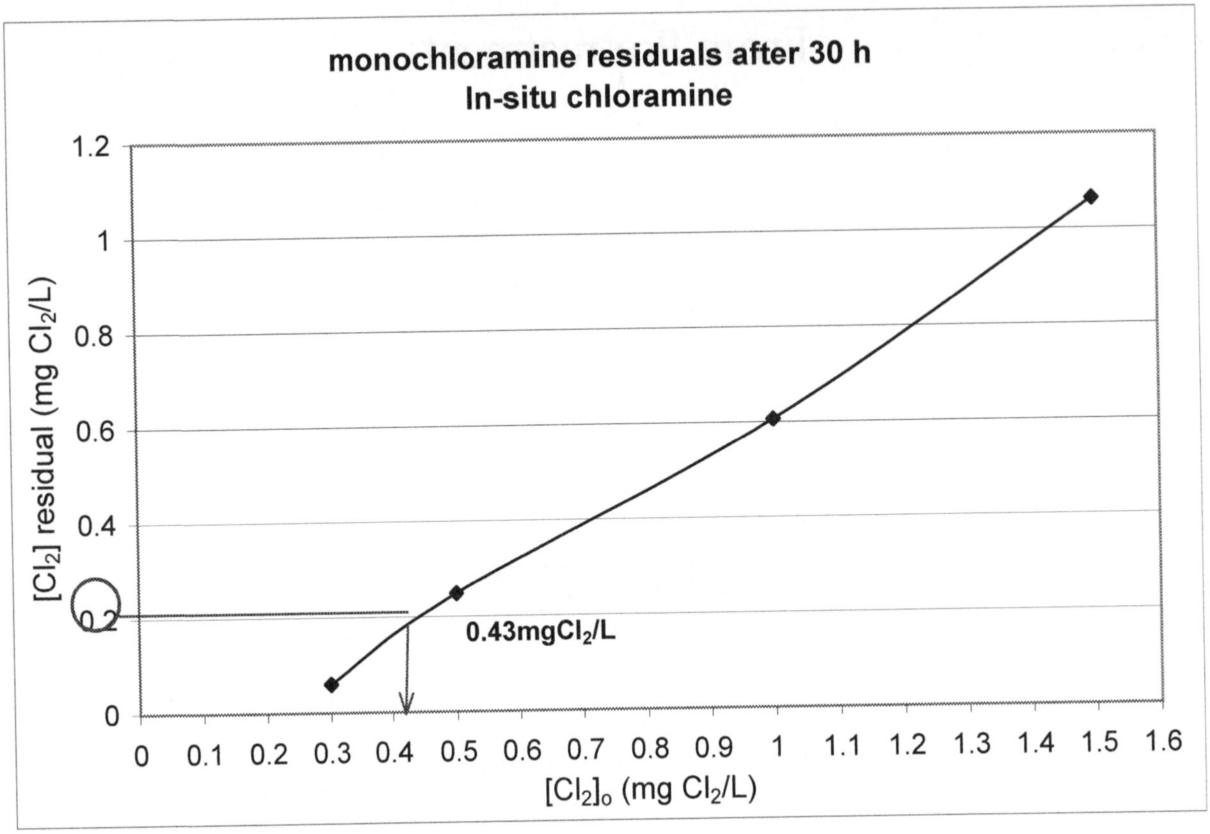

Figure 5.46 Chloramine (in situ) demand test on raw water

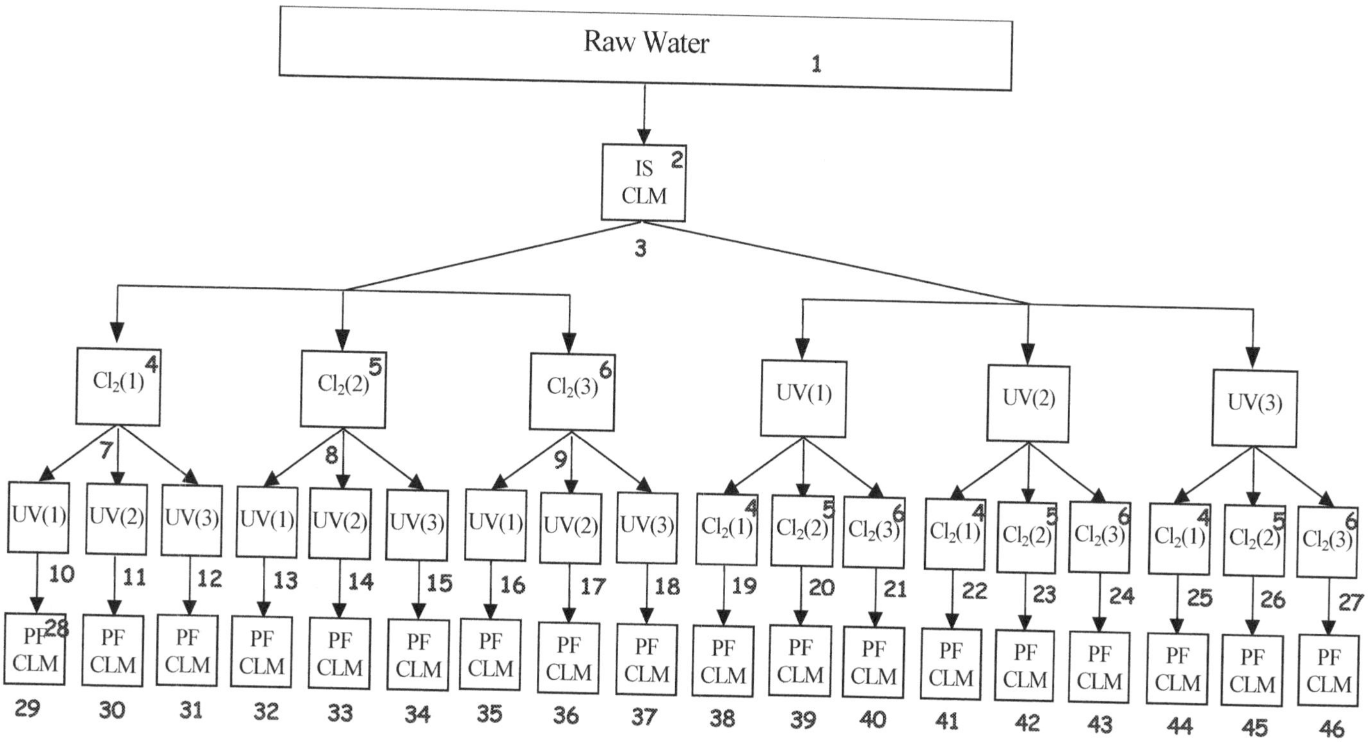

Figure 5.47 Sample numbering key for chlorine dose/residual data – DBP Test No. 8

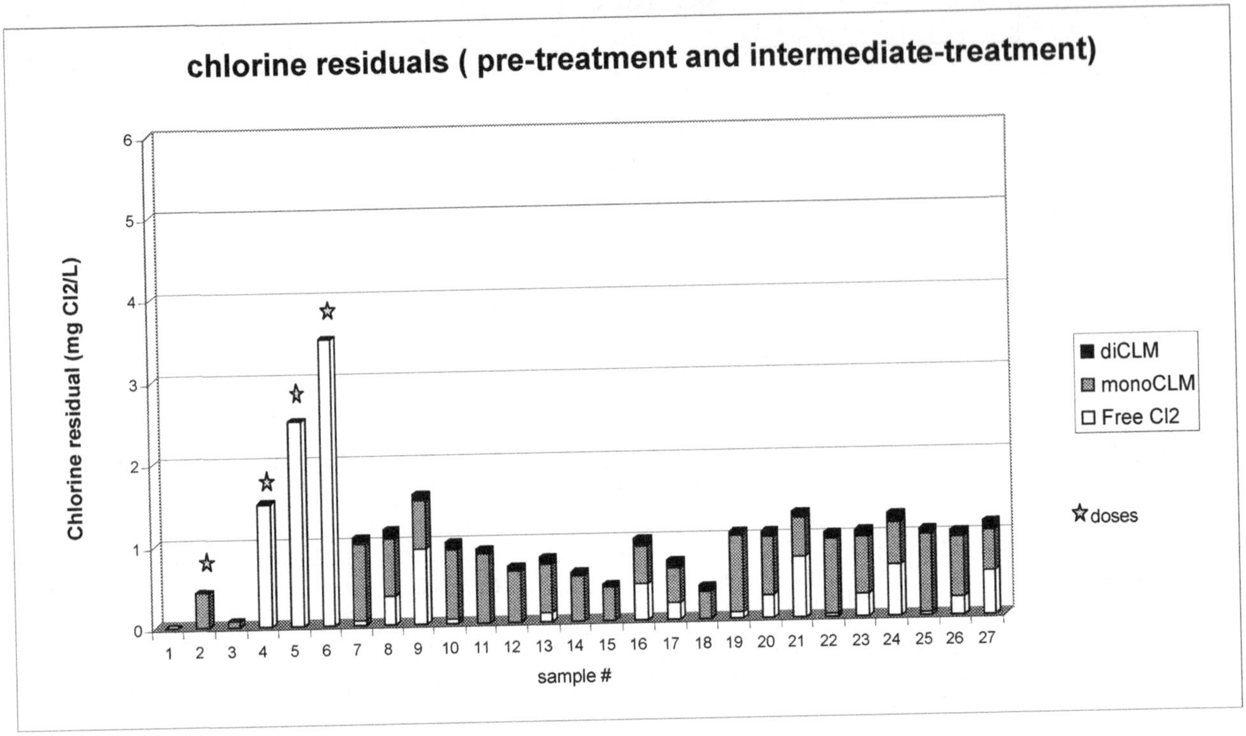

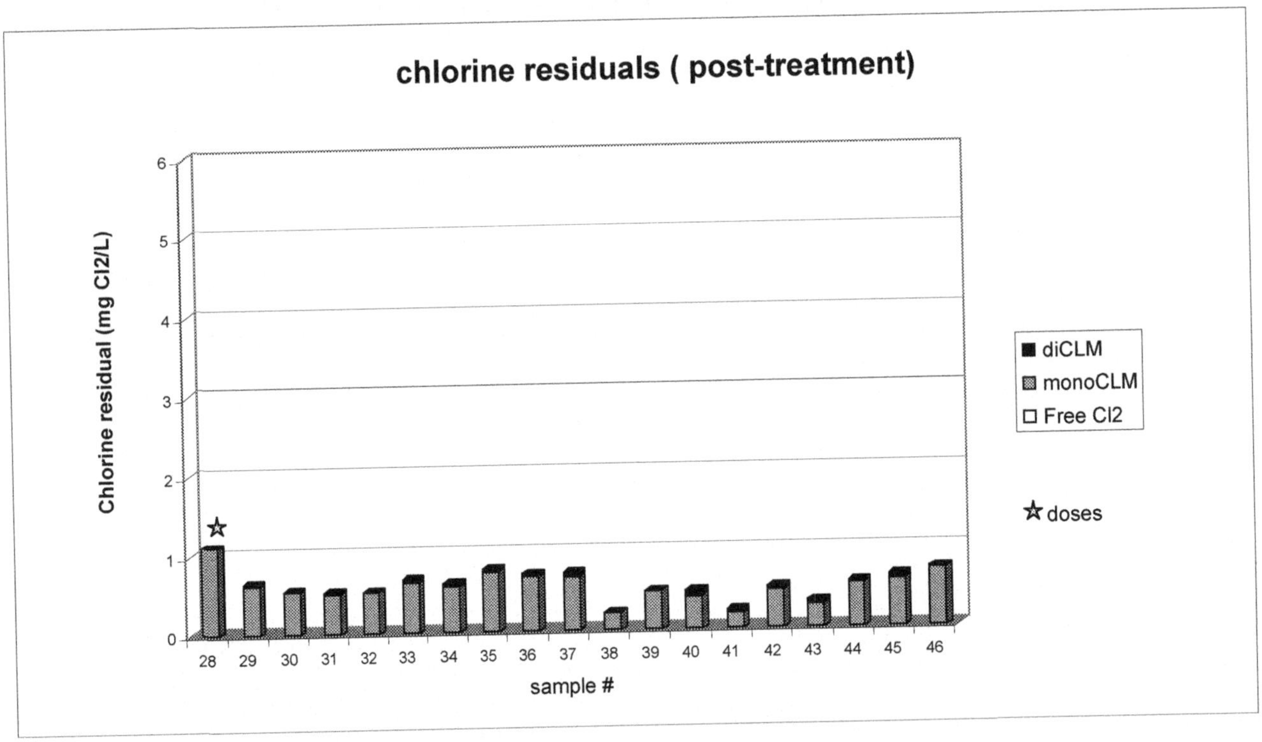

Figure 5.48 Chlorine dose and residuals DBP Test No. 8

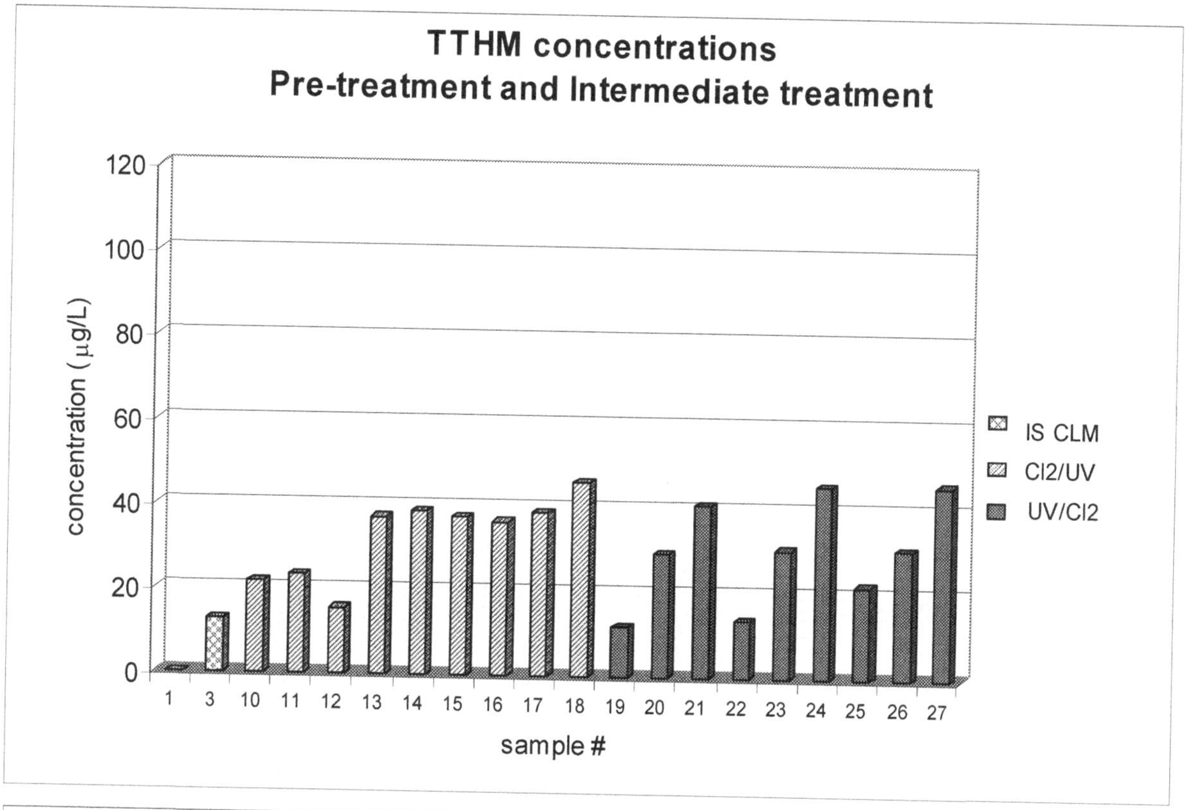

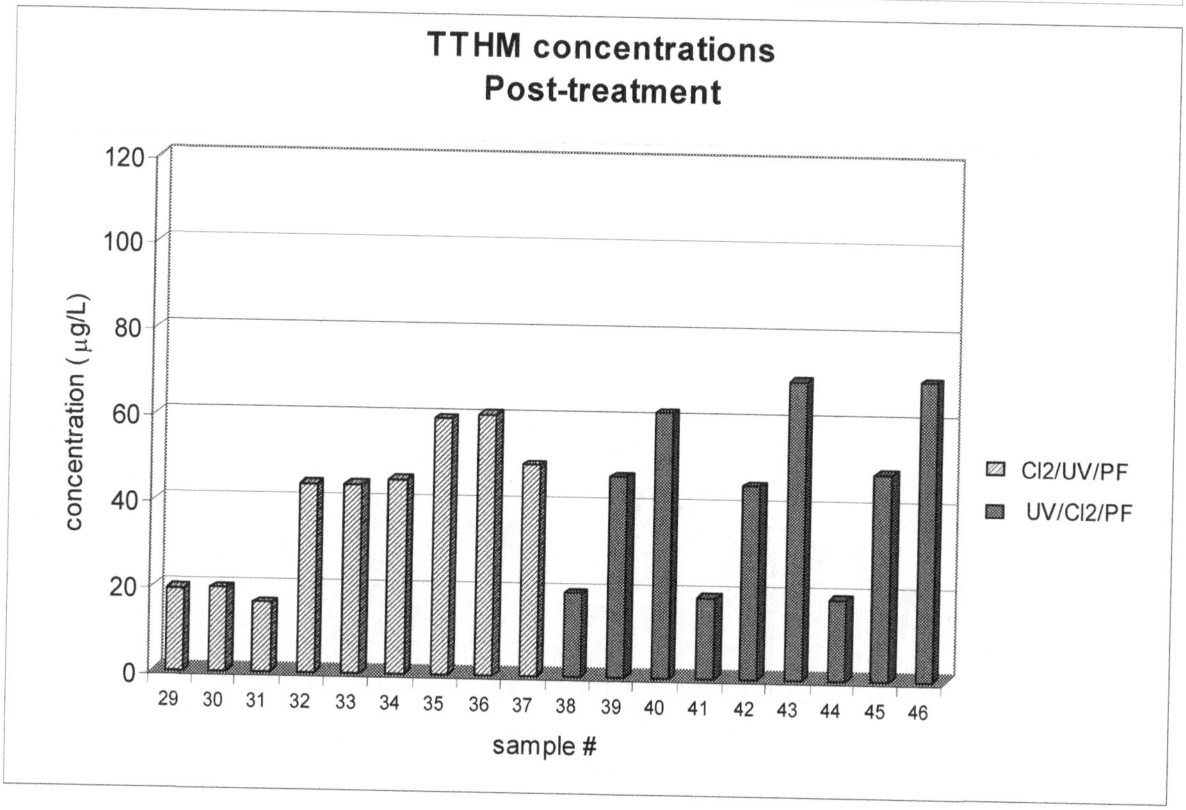

Figure 5.49 TTHM results from DBP Test No. 8

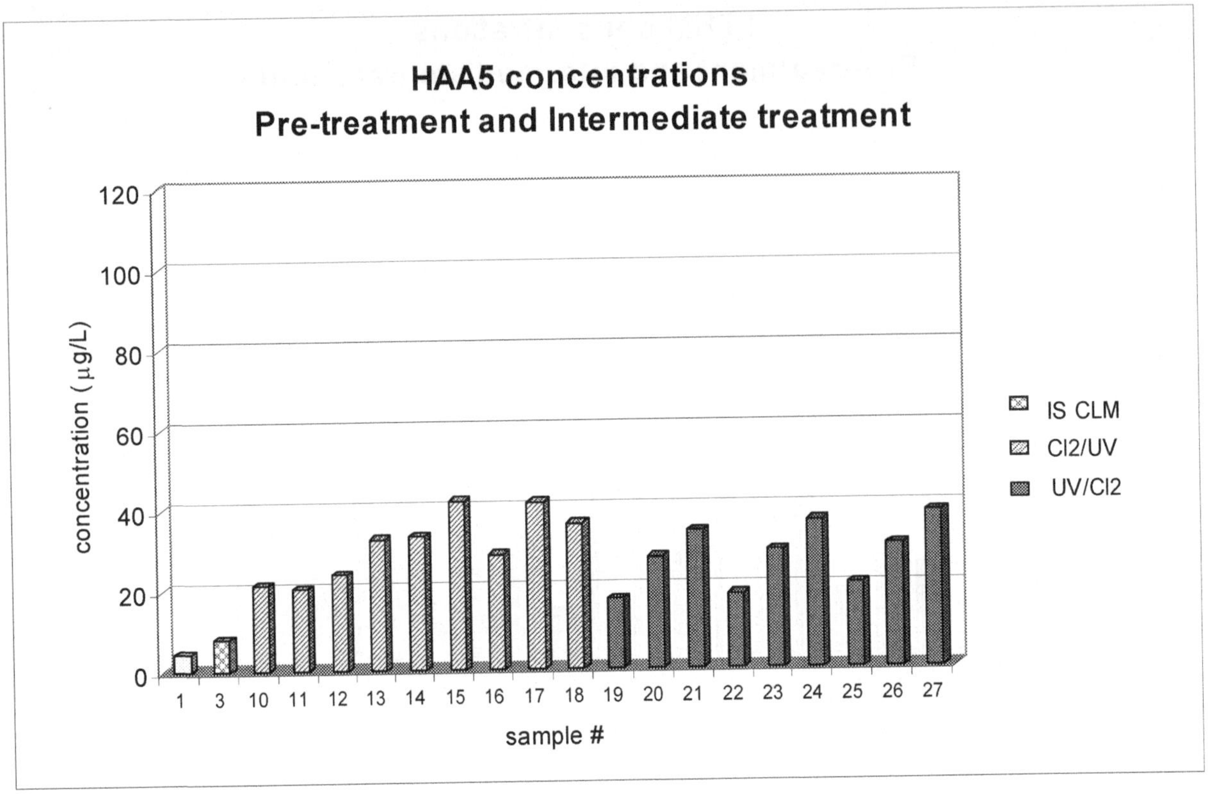

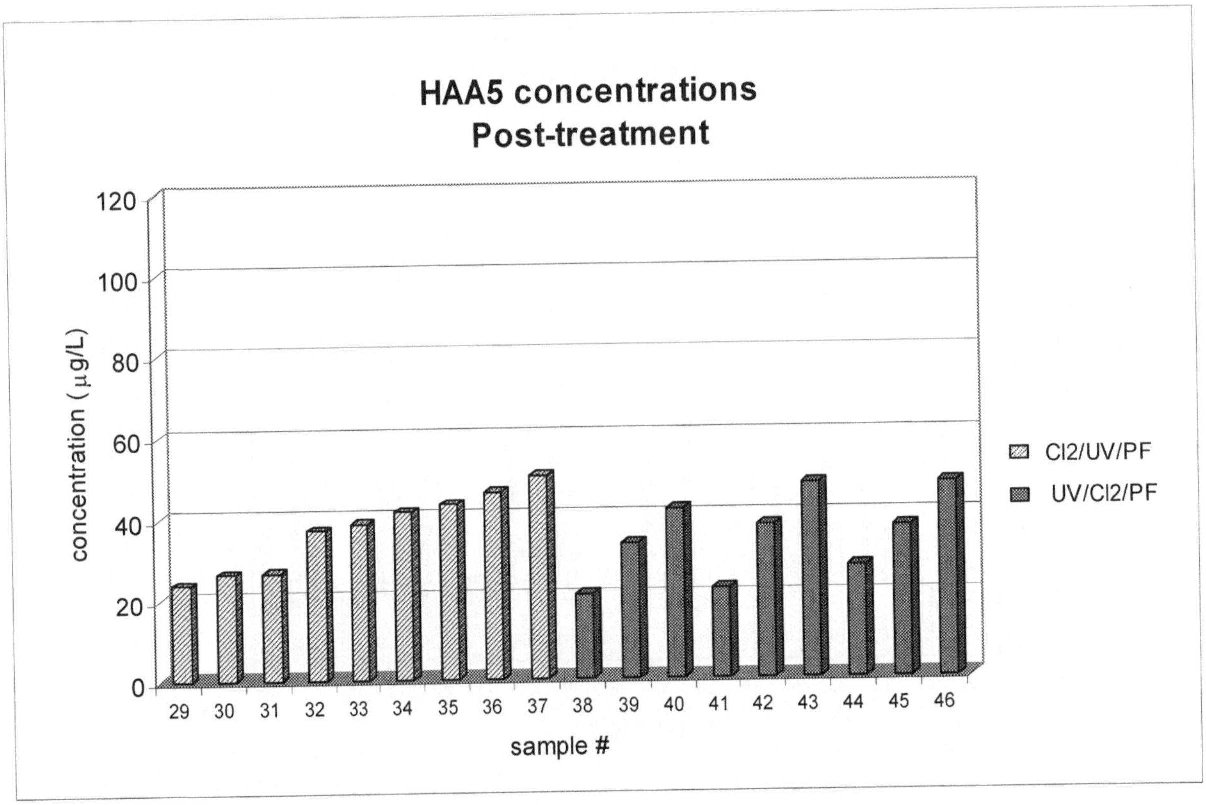

Figure 5.50 HAA results for DBP Test No. 8

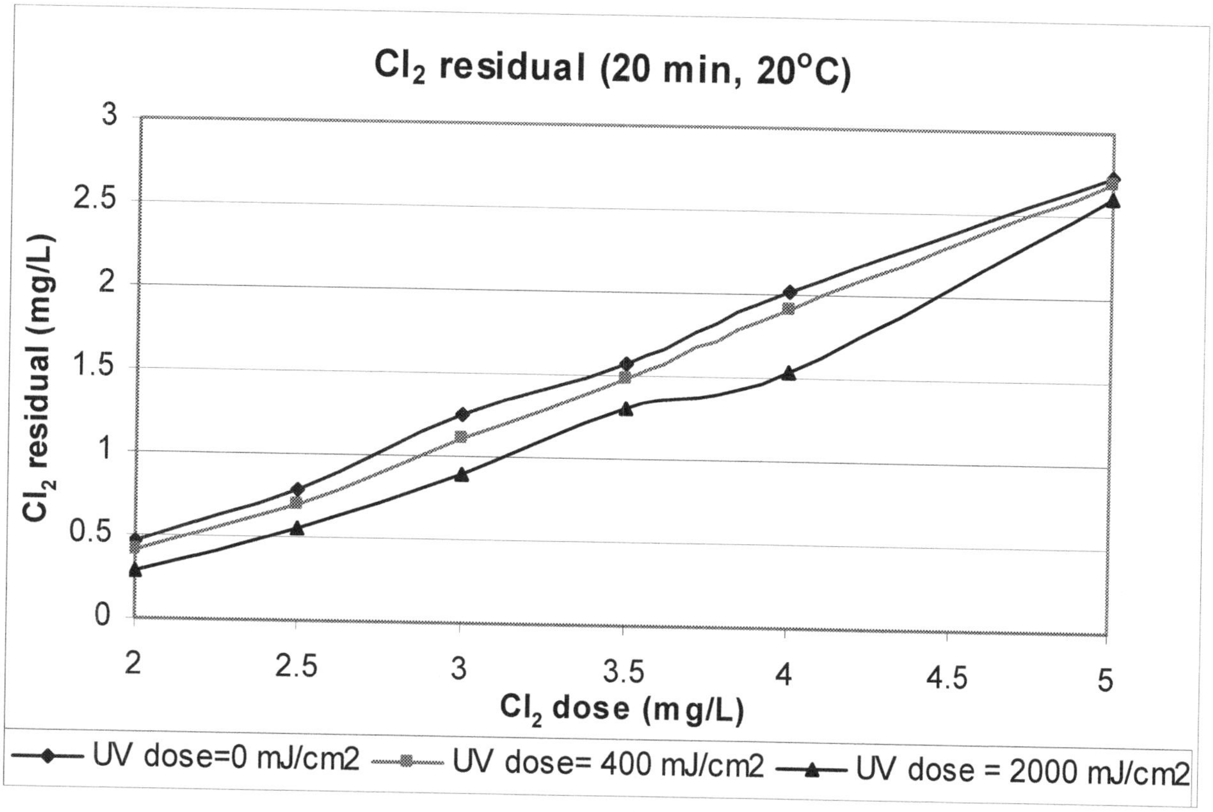

Figure 5.51 Cl_2 residuals after 20-minute incubation – DBP Test No. 8

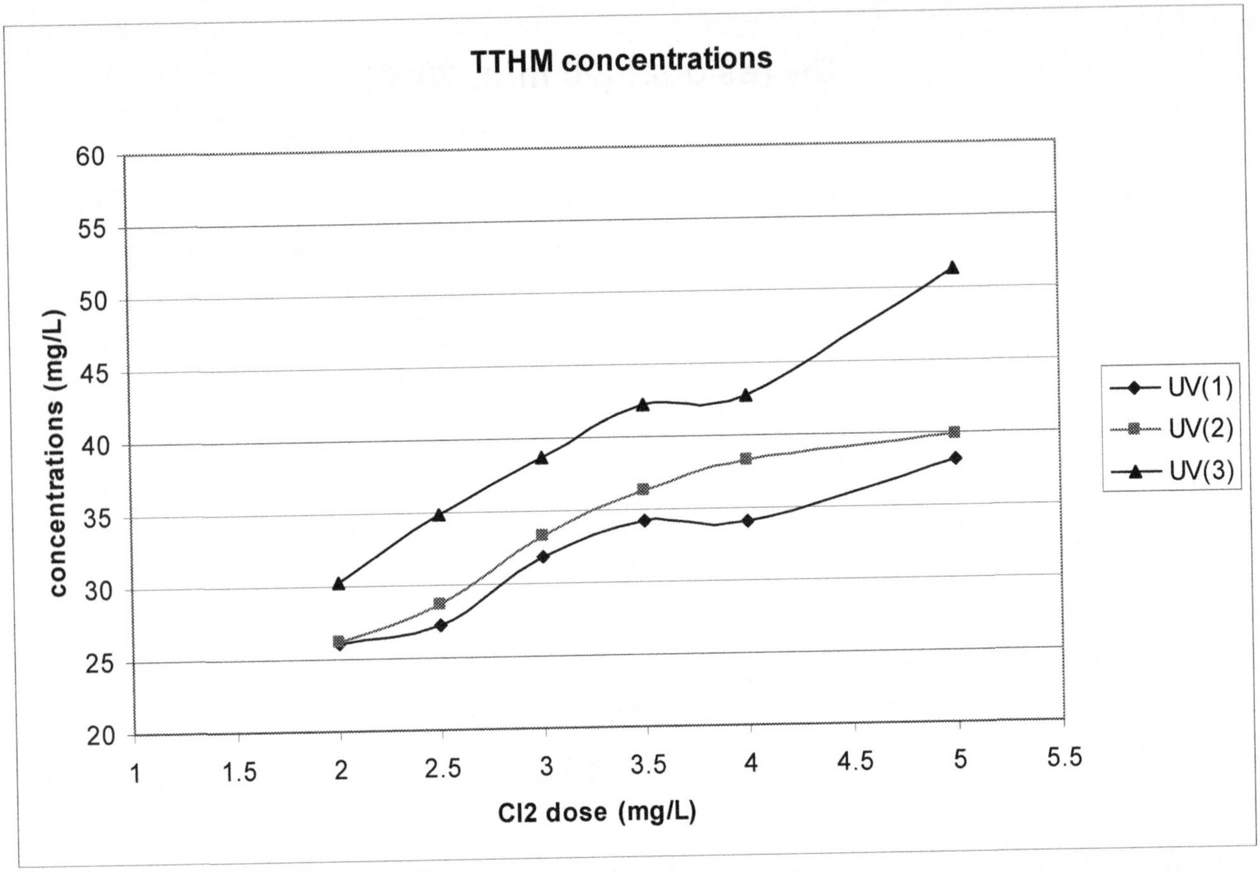

Figure 5.52 TTHM concentration profiles – DBP Test No. 8

CHAPTER 6
CONCLUSIONS AND RECOMMENDATIONS

SUMMARY OF RESEARCH FINDINGS

Inactivation Studies

The five inactivation (biodosimetry) studies showed that both low-pressure high output and medium-pressure UV reactors performed predictably on the unfiltered supply water. Inactivation and dose targets of approximately 40 mJ/cm^2 were achieved on ambient water quality conditions. Both reactors delivered the desired UV dose to the water through seasonal water quality variations.

The data also show the importance of routine lamp sleeve cleaning. Cleaning was necessary at approximately 1,500-hour intervals or less for the Winnipeg water quality.

Turbidity spiking events carried out during Inactivation Study Nos. 4 and 5 indicated that sediments from the source water basin had little to no effect on the RED and that no significant change in reactor performance was observed at turbidity levels up to 4 NTU. There was also no effect on the CB study (i.e., the ambient and 4 NTU water samples had the same RED results). At 13 NTU, there was a decrease in treatment efficiency. Because Winnipeg's unfiltered water rarely exceeds 2 NTU, turbidity should not be an issue with respect to reactor performance and disinfection efficacy. These results also confirm that UV disinfection can be used to meet the *Cryptosporidium* and *Giardia* disinfection requirements for unfiltered systems under current USEPA regulations.

The UV and organic test results show that increasing TOC results in decreasing UVT. This is expected as TOC will affect UV absorbance. During higher TOC events, UV reactors will require higher energy to maintain the desired dose. This effect on the UV reactors is expected to be minor due to the narrow range of TOC values (9 to 11 mg/L) typical for this source water. No TOC effect was seen on reactor performance and disinfection efficacy once the impact on UVT was accounted for.

The algae studies indicate that there is a consistent correlation between plankton counts and UVT. For varying plankton count levels, UVT values range between 74.6 and 77.1 percent. This suggests that UV reactors will require higher energy during elevated plankton levels in the Deacon Reservoir due to the decrease in UVT. However, the algae do not physically interfere with UV disinfection efficacy. The Shoal Lake sample microorganism response to CB testing was not significantly affected by the presence of the plankton. Once the UV intensity was corrected, the MS-2 had the same response to delivered UV dose as the samples where there are no high plankton counts. This suggests the plankton is not shielding or otherwise protecting the MS-2 biodosimeter from inactivation by the UV light. This conclusion is further supported by the limited impact that turbidity had on UV disinfection for waters with less than 4 NTU.

Operational Considerations

UV intensity sensor reliability needs to be addressed in the selection and operation of full-scale UV systems. Both pilot test reactors experienced difficulties with online intensity sensors.

The Aquionics UV intensity sensor was extremely sensitive to turbidity fluctuations as observed during the turbidity spiking tests. However, in Challenge Test Nos. 2 through 4, the sensor failed to indicate any significant difference in performance pre- and postcleaning of the lamp sleeves, while the biodosimetry showed a significant change.

The Wedeco sensor output saw only gross changes in turbidity (0 to 4 to 13 NTU). This sensor indicates what the biodosimetry results confirm – that changes in turbidity have only a small effect on delivered dose. Yet, this sensor also failed during Challenges No. 3 and No. 4 to indicate any significant difference in performance pre- and postcleaning of the lamp sleeves.

UV intensity sensors can be a valuable tool in monitoring system performance; however, from the inactivation studies, it is clear that the sensors can fail to fully describe what is happening in the system. It is believed that inaccurate sensor results were, in part, due to the pilot-scale equipment and a series of operational difficulties encountered by both UV reactors during the pilot testing.

The most significant influence on reactor performance observed during this study resulted from cleaning the quartz sleeves. Relatively frequent cleaning of the sleeves appears to be the most desirable maintenance plan because the sensors do not fully account for loss in light intensity due to dirty sleeves. For this unfiltered source, it appears that the Aquionics lamp and Wedeco lamps can operate for approximately 2,000 hours before cleaning is required. Due to water quality fluctuations, the time between cleanings will be different for different water sources.

From June 5, 2001, through July 29, 2002, the Aquionics unit exhibited a 3.38 kW average power draw for given conditions of transmittance and to achieve the target dose. This corresponds to 9.02 kW/mgd at an average UVT of 77 percent.

From June 5, 2001, through July 29, 2002, the Wedeco unit exhibited a 0.61 kW average power draw for given conditions of transmittance and to achieve the target dose. This corresponds to 5.0 kW/mgd at an average UVT of 77 percent.

Disinfection Byproduct Tests

Disinfection Byproduct Test Nos. 1 Through 5

THMs. The various disinfection scenarios resulted in THM concentrations between 25 and 55 µg/L for those with prechloramination and 45 to 120 µg/L for those with prechlorination. THM results and trends were consistent for DBP Test Nos.1 through 5.

HAA5. Haloacetic acid levels in the final treated samples ranged from 25 to 55 µg/L when prechloramination was used. The corresponding values for prechlorination trains were 40–100 µg/L. The HAA results and trends were consistent for DBP Test Nos. 1 through 5.

Aldehydes and Ketoacids. Aldehyde data showed a substantial increase due to the application of chlorine-based disinfectants. Final aldehyde levels were in the range of 25 to 55 µg/L. These aldehyde levels were similar to the raw water aldehyde levels seen during the 1997 pilot program.

Ketoacid levels were only moderately elevated. Final ketoacid levels were in the range of 10 to 40 µg/L. There seemed to be very little difference for nonchlorinated byproducts formed when comparing chlorination versus chloramination.

Chemical Disinfection Process Sequencing

Data from DBP Test Nos. 1 through 5 showed no obvious difference between the sequencing of the UV and chlorine. Although some tests showed the sequence of Cl_2/UV to offer a modest advantage in controlling THM and HAA concentrations, other tests showed the reverse. Furthermore, the benefits in using one sequence versus the other were usually dampened or completely lost when viewing the final chloraminated DBP data.

AOC and BDOC data also indicate that the sequencing of chlorine and UV is not important.

There was no clear indication that preformed chloramines led to reduced levels of DBPs as compared to in situ formed chloramines.

DBP Test Nos. 6 Through 8

DBP Test No. 6 indicated that high UV doses caused a small increase in chlorine demand. At a UV dose of 400 mJ/cm^2, the increase in free-chlorine demand was approximately 20 percent. The increase in demand was primarily the result of the UV destruction of chlorine. This increase in chlorine demand is expected to be minor at typical UV doses used for disinfection in water treatment (i.e., <60 mJ/cm^2). In general, at high UV dose levels (400 to 2,000 mJ/cm^2), DBPs also increased with increasing UV dose. In some cases, these increases were as high as 20 percent over the control. This increase in DBPs is expected to be minor at typical UV doses used for disinfection in water treatment (i.e., <60 mJ/cm^2). This test showed a trend indicating lower DBP formation in the UV/Cl_2 sequence as compared to the reverse (Cl_2/UV) at high UV doses.

DBP Test No. 7 indicated that the partial DBP modeling yielded the following equations for THMs and HAA5:

$$\text{THM } (\mu g/L) = 10^{-0.021} + (Cl_2 \text{ dose } [mg/L])^{0.444} + (\text{Time } [hrs.])^{0.191} + (pH)^{1.518}$$
$$\text{HAA5 } (\mu g/L) = 10^{1.771} + (Cl_2 \text{ dose } [mg/L])^{0.718} + (\text{Time } [hrs.])^{0.128} + (pH)^{-0.578}$$

DBP Test No. 7 showed that DBP formation increases with chlorine dose and contact time under all pH conditions. It also showed that THM concentration increased with pH, while HAA5 concentration decreased with pH at all doses and contact times. The general trends for the kinetic data for this water were generally in agreement with those reported for other waters in the literature.

DBP Test No. 8 indicated that there was no obvious difference between the sequences UV/Cl_2 and Cl_2/UV with respect to the THM and HAA formation. However, at high UV doses (400 to 2,000 mJ/cm^2), this test showed a clear trend indicating lower TOX and NDMA formation in the Cl_2/UV sequence as compared to the reverse (UV/Cl_2). The increase in TOX concentration in the UV/Cl_2 sequence was as much as 44 percent. For a single sample set, the NDMA concentration of the UV/Cl_2 sequence was almost doubled compared to the Cl_2 /UV sequence at a UV dose of 400 mJ/cm^2. As with DBP Test No. 6, this test also confirmed that UV treatment caused a noticeable increase in chlorine demand at high UV doses. DBP concentrations also increased substantially with increasing UV dose resulting from a corresponding increase in applied chlorine. This was especially noticeable with TOX concentrations, which increased with increasing UV dose. This increase was as high as

34 percent when the UV dose went from 0 mJ/cm^2 to 400 mJ/cm^2 and 62 percent when the dose was increased to 2,000 mJ/cm^2.

MAJOR FINDINGS

For this unfiltered water supply, both low- and medium-pressure UV reactors delivered the desired 40 mJ/cm^2 UV dose to the water through seasonal water quality variations. Dose determinations were best made by the biodosimetry testing.

Turbidity spiking indicated that the sediments from the source water basin had little to no effect on the RED and that no significant change in reactor performance was seen at turbidity levels up to 4 NTU. At a turbidity level of 13 NTU, there was a reduction in UV disinfection efficacy. Because Winnipeg's unfiltered water rarely exceeds 2 NTU, turbidity should not be an issue with respect to reactor performance and disinfection efficacy. Also, for other unfiltered supplies that meet USEPA unfiltered water requirements, UV can provide the required disinfection without compensation for turbidity.

Increasing TOC results in decreasing UVT. During higher TOC events, UV reactors will require higher energy to maintain the desired dose due to the reduction in UVT. Higher TOC values were seen as a result of increased algal concentrations and natural changes in the source water matrix. This effect on the UV reactors is expected to be minor with source waters with a narrow range of TOC values. No TOC effect was seen on reactor performance and disinfection efficacy once the lamp intensity was corrected for the decrease in UVT resulting from the increase in TOC.

There is an inverse relationship between plankton counts and UVT. This suggests that UV reactors will require higher energy during elevated plankton levels in source waters. Once the UV intensity was corrected, the MS-2 had the same response to delivered UV dose as the samples where the plankton counts were not high. This confirms that the plankton are not shielding or otherwise protecting the MS-2 biodosimeter from UV inactivation.

UV intensity sensor performance bears watching in full-scale UV system selection, design, and operation. It is essential during UV reactor validation and long-term operation that the sensors meet tight specifications initially and are proven periodically to be calibrated and responsive to significant changes in performance (RED) caused by such common factors as lamp fouling, flow changes, lamp aging, etc.

For this source water, UV reactor sleeves require cleaning approximately every 2,000 hours. This cleaning cycle was estimated based on the pre- and postcleaning inactivation study results. Lamps and sleeves require replacement after approximately 5,000 hours of operation. These cleaning and replacement periods will differ for varying source waters due to varying water quality.

TTHM levels as produced by the various disinfection scenarios resulted in concentrations between 25 and 55 μg/L for prechloramination scenarios and 45 to 120 μg/L for those involving prechlorination. Haloacetic acid levels in the final treated samples ranged from 25 to 55 μg/L when prechloramination was used. The corresponding values for prechlorination trains were 40 to 100 μg/L. With prechloramination, these levels should meet near-term DBP goals and regulations. In the long-term, the City will be implementing water treatment that will further reduce DBP precursors and resulting DBP levels will be lower.

DBP tests indicated that there was no obvious difference between the sequences UV/Cl_2 and Cl_2/UV with respect to the THM and HAA formation. At the high UV doses tested (400 to

2,000 mJ/cm^2), a clear trend indicated lower TOX and NDMA formation in the Cl_2/UV sequence as compared to the UV/Cl_2 sequence. Also at these high UV doses, DBP testing indicated that UV treatment caused a noticeable increase in chlorine demand, as expected, due to the destruction of chlorine by the UV. Between 0 and 400 mJ/cm^2, the increase in demand was about 20 percent. It is anticipated that UV treatment and UV/CL2 sequencing will have minor effects on DBPs and chlorine demand at UV doses typically used for water treatment.

RECOMMENDATIONS FOR FUTURE STUDY

At high UV doses, the impacts of UV dose on chlorine demand and DBP formation as indicated by DBP Test No. 8 are important and should be investigated further. Between UV doses of 0 and 400 mJ/cm^2, the increase in chlorine demand was about 20 percent. DBP concentrations also increased substantially with increasing UV dose. TOX concentrations increased with increasing UV dose by as much as 34 percent when the UV dose went from 0 mJ/cm^2 to 400 mJ/cm^2.

The sequence of chlorine and UV was also important. At high UV doses, the increase in TOX concentration in the UV/Cl_2 sequence was as much as 44 percent higher than the Cl_2/UV sequence. For a single sample set, the NDMA concentration of the UV/Cl_2 sequence was almost doubled compared to the Cl_2/UV sequence. These results may not be a concern to utilities that are planning to operate a system using the UV/Cl_2 sequence because the UV doses typically used in water treatment should be at much lower levels. However, some testing should be carried out to confirm that UV impacts on DBPs and chlorine demand are minor in the UV dose range of 20 to 60 mJ/cm^2.

Two additional areas of investigation that would be beneficial include (1) some further quantification of UV dose targets, corresponding free-chlorine dose and contact time requirements for virus inactivation, and expected distribution system DBPs under various scenarios and (2) further sensor research to find a sensor(s) and sensor locations that can accurately account for fluctuations in turbidity and organic levels in the source water, as well as lamp fouling.

RECOMMENDATIONS TO UTILITIES

UV intensity sensor performance bears watching in full-scale UV system selection, design, and operation. It is essential during UV reactor validation and long-term operation that the sensors meet tight specifications initially and are proven periodically to be calibrated and responsive to significant changes in performance (RED) caused by such common factors as lamp fouling, flow changes, and lamp aging.

Utilities should investigate the impacts of UV dose on chlorine demand and DBP formation for their water and the impacts of sequencing of UV and free chlorine on DBP formation.

It is reasonable to consider UV disinfection for unfiltered systems that comply with the USEPA regulatory requirements.

REFERENCES

Adams, M.H. 1959. *Bacteriophages*. New York: Interscience Publishers, Inc.

Allgeier, S.C., R.S. Summers, J.G. Jacangelo, V.A. Hatcher, D.M. Moll, S.M. Hooper, J.W. Swertfeger, and R.B. Green. 1996. A Simplified and Rapid Method for Biodegradable Dissolved Organic Carbon Measurement. Proceedings of the Water Quality Technology Conference, Denver, CO, American Water Works Association.

Amy, G.L., M. Siddiqui, K. Ozekin, X. Wang, H.W. Zhu, and J. Westrick. 1993. Empirically Based Models for Predicting Chlorination and Ozonation By-Products: Haloacetic Acids, Chloral Hydrate, and Bromate. Annual Conference Proceedings, American Water Works Association, Water Research, San Antonio, TX, 113–127.

Amy, G.L., P.A. Chadik, and Z.K. Chowdhury. 1987. Developing Models for Predicting Trihalomethane Formation Potential and Kinetics. *Jour. AWWA*, 79(7): 89–97.

Carlson, K.H., and G.L. Amy. 2000. The Importance of Soluble Microbial Products (Smps) in Biological Drinking Water Treatment. *Wat. Res.*, 34(4): 1386–1396.

Christensen, J., and K.G. Linden. 2001. Ultraviolet Disinfection of Unfiltered Drinking Water: Particle Impacts. Proceedings of the IUVA First International Congress on Ultraviolet Technologies, Washington, D.C., June 14–16.

Chowdhury, Z.K., K. Dickerson, and G.L. Amy. 1994. Modeling the Formation of Haloacetic Acids and Trihalomethanes in Chemically Coagulated Waters. Proceedings of 1993 Water Quality Technology Conference, Part I, Miami, FL, 493–512.

Escobar, I.C., and A.A. Randall. 2001. Assimilable Organic Carbon (AOC) and Biodegradable Dissolved Organic Carbon (BDOC): Complementary Measurements. *Wat. Res.*, 35(18): 4444–4454.

Fair, Pat. 2000. Personal Communication.

Frias, J., F. Ribas, and F. Lucena. 1995. Comparison of Methods for the Measurement of Biodegradable Organic Carbon and Assimilable Organic Carbon in Water. *Wat. Res.*, 29(12): 2785–2788.

Garcia-Araya, J. F., J.P. Croué, F.J. Beltrán, and B. Legube. 1995. Origin and Conditions of Ketoacid Formation During Ozonation of Natural Organic Matter in Water. *Ozone Sci. Eng.*, 17(6): 647–656.

Huck, P.M., P.M Fedorak, and W.B. Anderson. 1990. Methods for Determining Assimilable Organic Carbon and Some Factors Affecting the Van Der Kooij Method. *Ozone Sci. Eng.*, 12(4): 377–392.

Kaplan, L.A., T.L. Bott, and D.J. Reasoner. 1993. Evaluation and Simplification of the Assimilable Organic Carbon Nutrient Bioassay for Bacterial Growth in Drinking Water. *App. Envir. Microbio.*, 59(5): 1532–1539.

Malley, J.P., Jr., and L.B. Passantino. 2001. The Effect of Turbidity and Algae on Dose Requirements for Ultraviolet Systems Treating Unfiltered Drinking Water Supplies. Proceedings of the IUVA First International Congress on Ultraviolet Technologies, Washington, D.C., June 14–16.

Paode, R.D., G.L. Amy, S.W. Krasner, R.S. Summers, and E.W. Rice. 1997. Predicting the Formation of Aldehydes and Bom. *Jour. AWWA,* 89(6): 79–93.

Reckhow, D.A., P.S. Rees, and K.R.L. Nusslein. 2003. *Quality Assurance Project Plan for Watershed Sources and Long Term Variability of BDOM and NOM as Precursors.* Denver, CO: AWWARF.

Schechter, D.S., and P.C. Singer. 1995. Formation of Aldehydes During Ozonation. *Ozone Sci. Eng.*, 17(1): 53–69.

Schmidt, Wido, Wricke, Burkhart, Böhme, Ute, and W. Baumgardt. 1999. Impact of Ozonation on Formation of Brominated Organics. In *Proc. of the 14th Ozone World Congress,* Volume 1, 367–383. Stamford, CT: International Ozone Association.

Volk, C. et al. 1994. Comparison of Two Techniques for Measuring Biodegradable Dissolved Organic Carbon in Water. *Envir. Technol.,* 15, 545.

Volk, C.J., and M.W. Lechevallier. 2002. Effects of Conventional Treatment on Aoc and Bdoc Levels. *Jour. AWWA,* 94(6): 112–123.

Xie, Y., and D.A. Reckhow. 1992. A New Class of Ozonation By-Products: The Ketoacids. 1992 Annual Conference Proceedings, American Water Works Association. *Wat. Qual.*, 251-265.

Xie, Y., and D.A. Reckhow. 1992. Identification and Quantification of Ozonation Byproducts: Ketoacids in Drinking Water. Proceedings of the IOA Pan American Conference: International Ozone Association.

Xie, Y., and D.A. Reckhow. 1997. Development of a Method to Identify Ketoacids in Ozonated Fulvic Acid Solutions. *J. Org. Mass Spectrom.,* 32, 99–102.

ABBREVIATIONS

°C	degree Celsius
A	absorbance of the sample
AOC	assimilable organic carbon
AWWA	American Water Works Association
BDOC	biodegradable dissolved organic carbon
BG	blue-green
BOD	biological oxygen demand
BOM	biodegradable organic matter
CB	collimated beam
cm	centimeter
CLM	chloramine
CLR	calcium, lime, and rust
CRC	combined-residual chlorine
CT	residual concentration x contact time
CU	color units
D	diatoms
DBP	disinfection byproduct
DCAA	dichloroacetic acid
diCLM	dichloramine
DOC	dissolved organic carbon
DPD	method of measuring chlorine residual in water
ECD	electron capture detection
FRC	free-residual chlorine
GC	gas chromatograph
g/g	gram per gram
gpm	gallons per minute
HAA	haloacetic acid
HAA5	sum of 5 haloacetic acids
IS-CLM	in situ formed chloramines
IUVA	International Ultraviolet Association
km	kilometer
kW/MG	kilowatt per million gallon
kW/mgd	kilowatt per million gallons per day

kW-hr.	kilowatt-hour
L	liter
L/mg/m	liter per milligram per liter
LPHO	low pressure high output
LSB	laboratory solvent blank
LT2ESWTR	Long Term 2 Enhanced Surface Water Treatment Rule
M/M	molar per molar
mg/L	milligrams per liter
mJ/cm^2	millijoules per squared centimeter
mL	milliliter
mL/min.	milliliter per minute
MCL	maximum contaminant level
MP	medium pressure
MRDL	maximum residual disinfectant level
monoCLM	Monochloramine
NDMA	nitrosodimethylamine
nm	nanometer
No.	number
NOM	natural organic matter
NTU	nephalometric turbidity unit
O&M	operations and maintenance
PF	pigmented flagellate
PF-CLM	preformed chloramines
pfu/mL	perfusion time per milliliter
QA/QC	quality assurance/quality control
R^2	statistical correlation
RED	reduction equivalent dose
RO	reverse osmosis
SDS	simulated distribution system
sec.	second
SEM	scanning electron microscopy
SOC	synthetic organic chemicals
SOP	standard operating procedure
std.	standard
SUVA	specific ultraviolet absorbance
TFTC	too few to count
THAA	total haloacetic acids

THM	trihalomethane
TNTC	too numerous to count
TOC	total organic carbon
TTHM	total trihalomethanes
TOX	total organic halides
UMass	University of Massachusetts
UNH	University of New Hampshire
USEPA	U.S. Environmental Protection Agency
UV	ultraviolet
UVDGM	UV Design Guidance Manual
UVT	ultraviolet transmittance
w/gpm	watts per gallon per minute

IWA Publishing's authorised EU representative for General Product Safety Regulations is Diane D'Arras, 15 rue Duret, 75116 Paris, France, e-mail: safety@iwap.co.uk.

Printed and bound by CPI Group (UK) Ltd, Croydon, CR0 4YY
17/09/2026
02218804-0020